CONSTANT TOUCH

CONSTANT TOUCH

A Global History of the Mobile Phone

JON AGAR

ICON

First published in the UK in 2003 by Icon Books Ltd

This revised and updated edition published in the UK in 2013 by
Icon Books Ltd, Omnibus Business Centre,
39–41 North Road, London N7 9DP
email: info@iconbooks.net
www.iconbooks.net

Sold in the UK, Europe and Asia
by Faber & Faber Ltd, Bloomsbury House,
74–77 Great Russell Street, London WC1B 3DA

Distributed in the UK, Europe and Asia
by TBS Ltd, TBS Distribution Centre, Colchester Road,
Frating Green, Colchester CO7 7DW

Distributed in South Africa
by Book Promotions, Office B4, The District,
41 Sir Lowry Road, Woodstock 7925

Distributed in Australia and New Zealand
by Allen & Unwin Pty Ltd,
PO Box 8500, 83 Alexander Street,
Crows Nest, NSW 2065

Distributed to the trade in the USA
by Consortium Book Sales and Distribution,
The Keg House, 34 Thirteenth Avenue NE, Suite 101,
Minneapolis, MN 55413-1007

Distributed in Canada by
Penguin Books Canada,
90 Eglinton Avenue East, Suite 700,
Toronto, Ontario M4P 2Y3

ISBN: 978-184831-507-5

Typeset in Utopia by Marie Doherty

Printed and bound in the UK by
CPI Group (UK) Ltd, Croydon CR0 4YY

To Kathryn, Max and Hal (and his new Nokia 113)

Contents

Contents

About the author

Jon Agar is currently Senior Lecturer in the Department of Science and Technology Studies (STS) at University College, London. He has taught history of science and technology at the universities of Manchester, Cambridge and Harvard.

Acknowledgements

For the first edition of this book, many thanks to: Cath Skinner, Jon Turney and Simon Flynn for their comments on an early draft of *Constant Touch*; the staff at the British Library and the City Business Library for their help in locating documents; Mark Tindley for some excellent impromptu research on health matters (and to the rest of the Thursday night football crowd for their 'encouragement' to finish); and Brian Balmer for a good anecdote.

In preparation of this edition, my thanks to Simon Rockman (for corrections in Amazon review), to Imogen Clarke, Leucha Veneer, Jakob Whitfield and Charlotte Connelly (for kindly responding to my @jon_agar twitter plea for new movies featuring mobiles) and to Duncan Heath and Robert Sharman at Icon.

Preface

Some books start with a small observation that lodges in the memory and niggles away until the author feels compelled to stop what they are doing and devote time and thought to understanding what they have seen. *Constant Touch* is a case in point. In the 1990s my friends started carrying around mobile phones. I was a late adopter. By the time I got my first phone, its purchase was not so much a fashion statement or the result of any love of shiny new technology as it was a necessity: if I wanted a social life then I had better get my hands on one. My friends had long dispensed with the traditional method of arranging nights out well in advance, establishing firm details such as where and precisely when we would meet. Instead, a vague indication of plans would be finalised at the last minute, coordinated in real time through texts and quick calls. I got a phone because I did not want to be left out.

I had no idea then that this mundane tool would form the subject of a book. Instead, the observation that started me down that path was made on a visit to the United States. I'm a historian of science and technology, and one of the enjoyable perks of academic life is the conference season: we meet up, we argue about interpreting the past, and swap gossip. Usually, American stories are particularly prominent in the history of modern technology. From the second half of the 19th century through the 20th century, we are used to seeing the major innovations, as well as the defining patterns of production and consumption, emerge in the United States. Think of the works of Thomas Edison, Henry Ford and Alexander Graham Bell, or even NASA or the companies of Silicon Valley. I was going to a conference in Baltimore, where I was going to talk about a minor, if fraught, episode in the Cold War when it became an urgent project to use the moon as a relay of radio signals. In my mind I was rehearsing my case study, reflecting too on how the Cold War had pushed forward certain technologies, not least in the United States. Then I noticed something unexpected. I had come from a city, London, where cellphones were everywhere – in hands, on café tables, or close by in coat pockets. Certainly it was a novel enough development that people still complained about having to overhear personal conversations in places where usually polite etiquette ruled. But even the complaints were fading. Whinging was drowned out by ring tones. But in Baltimore the sights and sounds were different. Hardly any cellphones could be seen. When they did appear the conversations – if they even merited the term – were in contrast too: shorter, less chatty. How odd, I thought.

This small observation led me to ponder two questions. The first was why the manner of mobile phone use seemed to vary from place to place. I soon found out that the 'style', to use technology historian Thomas P. Hughes's terminology, of mobile phone networks as technological systems was indeed different in different countries. Furthermore, there were variations in mobile phone 'culture', or the meanings and expectations we find in patterns of use, that were rich and curious. The second question concerned the United States: it was strange to find a new technology in which other parts of the world seemed to be taking the lead. Was this true? And if so, why?

The first edition of *Constant Touch*, published in 2003, explored and offered answers to these questions. It may have started with an observation about the here and now, but I believe that the patterns of contemporary life can only be understood in historical perspectives. Some of the variations in mobile phone culture across the world are therefore best analysed as the outcome of changes over time, changes in which the social, cultural, political and technological are all intimately entwined. Even by 2003 the global number of cellphone subscribers was passing a billion. This was also part of the appeal: here was a technology that was available to many, and was sometimes marketed by multinational corporations as universal, yet was being taken up in diverse and even divergent ways. The global popularity of the mobile phone was an opportunity for studying how technologies are made sense of in the world's societies and cultures.

I was pleased with the reception accorded to the first edition of *Constant Touch*. My academic peers seemed to like it. Its 'acute analysis and engaging prose' was picked out by either the *Times Literary Supplement* or the *London Review of Books* (I forget which). *Vogue* called it the 'definitive book on the subject' – a stamp of approval that will shock and dismay anyone who has seen the way I dress. And the *Newbury Weekly News* provided a headline, 'Dry History Made Palatable by Entertaining Style', that still makes me smile and wish I carried business cards.

But cellphones are a fast-moving technology and by 2012 I knew, and my publishers agreed, that a new edition was needed. So much had happened, not least the transformation of the mobile phone from being primarily a tool for the communication of speech to becoming, with the smartphone especially, a richly visual, truly personal computer. However, yet again, the seed of the new book started with a common-place observation. While the title 'Constant Touch' had seemed apt because it caught the culture of informal and continuous contact found in cities such as Manila or London (if not yet Baltimore), now the words seemed applicable and important in a new way. I was travel-ling from home to work. I live in Dalston in the east of the capital and work at University College London. Catching the train at Dalston Kingsland, I realised, with a jolt, that every single fellow commuter was thumbing a smartphone. Each was fully absorbed. There was no rustling of newspapers or buzz of conversation – not even the one-sided conversations of the old cellphone days that invited you, as an unwilling eavesdropper, to fill in the blanks. Here was 'constant touch' all right, but of a different kind. And I recognised it in myself.

This new, expanded edition of *Constant Touch* brings the story up to date, and in particular tries to make sense of this second observation. The mobile phone was always an intimate technology, but why did the smartphone become more intimate still? What has driven the design and the desire that are its hallmarks? One of the ways of answering this question, of course, is to pay attention to Apple, a company that was entirely absent from the first edition. The extraordinary and defining intervention by Steve Jobs' company is a compelling reminder that American innovation (albeit with Chinese production) can still spring surprises.

The book has four parts. First, I guide the reader through how a 'cell' phone works, what makes it different from older, landline telephones, and where the idea came from. It is a delicious coincidence that the report proposing the cellular idea carried the name 'D.H. Ring'. From Bell's telephone to Ring's cellphone? In Part two, I trace how the idea of the cellular phone was, after many years languishing, realised in the first (analogue) and second (digital) generations of mobile phones. The story, crucially, is different in different regions. Also striking is how the mobile phone developed in close connection to other technologies of mobility, especially the automobile. Part three explores mobile cultures, as well as many other dimensions. Finally, Part four examines the mutation of this technology and culture as it became the 'smartphone'.

Part one
World in bits

Chapter 1
What's in a phone?

You can tell what a culture values by what it has in its bags and pockets. Keys, combs and money tell us that property, personal appearance or trade matter. But when the object is expensive, a more significant investment has been made. In our day the mobile or cellphone is just such an object. But what of the past? In the 17th century, the pocket watch was a rarity, so much so that only the best horological collections of today can boast an example from that time. But if in the following century you had entered a bustling London coffee shop or Parisian salon, then you might well have spied a pocket watch amongst the breeches and frock-coats. The personal watch was baroque high technology, a compact, complex device that only the most skilful artisans could design and build. Their proud owners did not merely buy the ability to tell the time, but also bought into particular values: telling the time mattered to the entrepreneurs and factory owners who were busy. As commercial and industrial economies began to roar, busy-ness conveyed business – and its symbol was the pocket watch.

At first sight it might seem as if owning a pocket watch brought freedom from the town clock and the church bell, making the individual independent of political and religious authorities. Certainly, possession granted the owner powers over the watchless: power to say when the working day might begin and finish, for example. However, despite the fact that the pocket watch gave the owner personal access to the exact time, this very accuracy depended on being part of a *system*. If he was unwilling personally to make regular astronomical observations, his pocket watch would still have to be reset every now and then from the town clock. With the establishment of time zones, the system within which a pocket watch displayed the 'right' time spread over the entire globe. 7.00am in New York was *exactly* 12.00pm in London, which was *exactly* 8.00pm in Shanghai. What is more, the owner of a pocket watch could travel all day – could be mobile – and still always know what time it was. Such certainty was only possible because an immense amount of effort had put an infrastructure in place, and agreements had been hammered out about how the system should work. Only in societies where time meant money would this effort have been worth it.

Pocket watches provide the closest historical parallel to the remarkable rise of the mobile cellular phone in our own times. Pocket watches, for example, started as expensive status symbols, but by the 20th century most people in the West possessed one. When cellular phones were first marketed they cost the equivalent of a small car – and you needed the car to transport them, since they were so bulky. But in 2002, global subscriptions to cellular phone services passed 1 billion. By 2010 the number was 5 billion, and in 2012 it was predicted that mobile devices of all kinds would soon outnumber human beings. In many countries most people have a mobile phone. Like the pocket watch, the phone had made the leap from being a technology of the home or street to being a much rarer creature indeed: something carried everywhere, on the person, by anybody. So, if pocket watches resonated to the rhythm of industrial capitalism, what values do the ringtones of the mobile phone signify? What is it about humanity in the 21st-century world that has created a desire to be in constant touch? To answer this question, I began with a rather drastic step.

In this world of weightless information, there is nothing quite so satisfying as taking a hammer to a piece of technology. My old mobile phone, a Siemens S8, was a solid enough device, and it took some effort to take it apart. But in the interests of research I wanted to know what was inside. Now the various parts lie in front of me.

The battery came away first. It is about the size of a large postage stamp, and as heavy as a paperback book. But I can at least lift it, and this would not have been true of the early years. Let's take an example of mobile radio communication from far back. The entrepreneurial expat Italian inventor Guglielmo Marconi had hawked his new technology of wireless telegraphy – radio communication by Morse code – around London in 1900, but he focused his efforts on one main customer: the mighty Royal Navy. His pitch was simple. The Admiralty had invested many thousands of pounds in battleships which became incommunicative and blind as soon as fog descended. Wireless telegraphy provided new mechanical senses: to warn of maritime dangers and to organise the fleet. The Sea Lords were convinced and Marconi sealed the deal. The Navy installed 32 wireless sets aboard ships. Sixteen years later, during the Battle of Jutland, the purchase

Guglielmo Marconi.

would prove a wise one: listening stations detected the German fleet by picking up unusually heavy radio traffic, and the same radio technology enabled the British Grand Fleet to steam across the foggy North Sea to the Danish coast to engage its enemy in the Skagerrak.

Marconi wanted to sell wireless to the Admiralty because it took a vehicle the size and power of a battleship to carry it. Radio transmission in the 1900s had been achieved by creating bursts of sparks generated by immense electrical voltages. (The same principle is behind the crackling interference caused by lightning.) Giant voltages meant heavy batteries. The first mobile radio was restricted to behemoths. But a feature of the history of electrical technology has been continuous miniaturisation of components. Even before the First World War, the Swedish electrical engineer Lars Magnus Ericsson had demonstrated new possibilities for mobile communication.

The young Ericsson had trained as a smith and as a mining and railway engineer before becoming an apprentice under the telegraph-maker A.H. Öller in the early 1870s. He then studied abroad in Switzerland and Germany before setting up his own company in Stockholm in 1876, first to manufacture and repair telegraph apparatus, and later, following Alexander Graham Bell's invention, telephones. Ericsson's business boomed. However, he seems to have wearied of the commercial life early in the new century, and retired, backed by a healthy bank balance, to a comfortable life as a farmer. But in 1910 Ericsson, in the spirit of tinkering, built a telephone into his wife Hilda's car: the vehicle connected by wires and poles to the overhead telephone lines that had sprung up even in rural Sweden. Enough power for a telephone could be generated by cranking a handle, and, while Ericsson's mobile telephone was in a sense a mere toy, it did work.

At one level the story of the retired Swedish engineer-turned-farmer is trivial: no great industry of car-carried mobile telephones was founded on the experiment. But in many other ways it was significant. Ericsson's company, after many twists and turns, would supply much of the infrastructure for the cellular phone systems built in the late 20th century. More of that later. Secondly, the experiment happened in Sweden, and the

An Ericsson table telephone, c. 1900. (BT Archives)

Nordic countries have a remarkable prominence in the history of the mobile phone. Finally, Ericsson's vehicle showed that the technologies of communication could be fitted in an automobile, the first instance in a long and profound association between two technologies of mobility that have shaped our modern world.

Marconi's weighty wireless had to be carried by battleship. Early practical mobile phones were carried by cars, since there was room in the boot for the bulky equipment, as well as a car battery to power them. One of the most important factors permitting phones that can be carried in pockets and bags has been a series of remarkable advances in battery technology. As batteries have become more powerful, so they have provided more energy. Partly because improvements in battery design have been incremental, their role in technological change is often underestimated. The great Prussian physicist Walther Hermann Nernst, who later articulated the Third Law of Thermodynamics, had experimented in Göttingen in 1899 with nickel as a means of converting chemical energy into electrical energy. Built a century later, my disintegrated phone has a Ni-MH – Nickel Metal Hydride – battery; it is in one sense recognisably similar to Nernst's, but in another it is transformed: it is many, many times lighter and more efficient. Step-by-step, nickel batteries have got better. Continuous experimentation with other metals has revealed slight but significant improvements, so that, for example, the early-21st-century choice for mass-produced energy packs is between nickel- and lithium-based techniques. Gradual change can eventually trigger a profound revolution. Once batteries became powerful *and* portable, a Rubicon was crossed. Uncelebrated improvements in batteries, put into laptops, camcorders and cellphones, triggered our mobile world.

A similar story can be told of the other bits and pieces in front of me. The LCD or liquid crystal display, the grey panel on which I read my incoming call numbers or SMS messages on my old phone, is now commonplace in consumer electronics. The contradictory properties of liquid crystals – fluids that can paradoxically retain structure – had been noted in the 19th century by the Austrian botanist Friedrich Reinitzer. He had noticed that the organic solid cholesteryl benzoate seemed to have two melting points, and that between the lower and higher temperatures the liquid compound behaved oddly. But it was not until the 1960s that industrial laboratories, such as RCA's in America, began to find applications exploiting this behaviour. Again, incremental development followed. Liquid crystal displays don't produce light, they reflect light, which potentially saves energy, so changes in one component (displays) interacted with another (batteries). Much effort was needed to turn this advantage into a practical one. However, by the 1970s LCDs appeared in calculators and digital watches, replacing the red glow of light-emitting diodes.

LCDs were not essential ingredients of a cellphone (indeed my new smartphone has ditched this old display technology). We could keep in constant touch with a simple assemblage of the other bits and pieces found in the wreckage of my phone: aerials, microphones, loudspeakers and electronic circuitry. But improved screens are part of what makes a mobile phone more than a mere instrument of communication. We don't just talk. Without the screen, the extra aspects of the mobile phone – the games, YouTube, the address books, the text messaging – all the features which contribute to a rich mobile culture, involving manipulation of data as well as transmission of the voice, would not be possible.

If I had superhuman strength I could hammer my phone into constituent atoms. A new global politics can be found among the dust. Mobile phones depend on quite rare materials: for example, within every phone there are ten to twenty components called capacitors, which store electrical charges, and since the Second World War the best capacitors have been made using thin films of a metal called tantalum. On the commodities market in the early 1990s, capacitor-grade tantalum could usually be bought for $30 a pound, sourced from locations such as the Sons of Gwalia mines at Greenbushes and Wodgina in Western Australia, the world's best source of the element. But in the last years of the 20th century, as more and more people bought mobile phones, the demand for tantalum shot up. The price per pound rose to nearly $300 in 2000.

Tantalum, in the form of columbite-tantalite ('coltan' for short), can also be found in the anarchic northeast regions of the Democratic Republic of Congo, where over 10,000 civilians have died and 200,000 have been displaced since June 1999 in a civil war, fought partly over strategic mineral rights, between supporters of the deceased despot Laurent Kabila and Ugandan and Rwandan rebels. As the price of tantalum increased, the civil war intensified, funded by the profits of coltan export. The mobile phone manufacturers are distanced, however, from the conflict. Firms such as Nokia, Ericsson, Samsung and Motorola buy capacitors from separate manufacturers, who in turn buy raw material from intermediaries. On each exchange, the source of tantalum becomes more deniable. 'All you can do is ask, and if they say no, we believe it,' Outi Mikkonen, communications manager for environmental affairs at Nokia, recently said of her firm's suppliers. On the other hand, export of tantalum from Uganda and Rwanda has multiplied twentyfold in the period of civil war, and the element is going somewhere.

To build a single cellphone requires material resources from across the globe. The tantalum in the capacitors might come from Australia or the Congo. The nickel in my battery probably originated from a mine in Chile. The microprocessor chips and circuitry may be from North America. The plastic casing and the liquid in the liquid crystal display were manufactured from petroleum products, from the Gulf, Texas, Russia or the North Sea, and moulded into shape in Taiwan. The collected components would have been assembled in factories dotted around the world. While the work might be coordinated from a corporate headquarters – Ericsson's base is in Sweden, Nokia's in Finland, Siemens' in Germany, Alcatel's in France, Samsung's in Korea, Apple's and Motorola's in the United States, and Sony's, Toshiba's and Matsushita's in Japan – the finished phone could have come from secondary manufacturers in many other countries.

The phone might be an international conglomerate, but it was put together in different ways in different countries, and shortly we will see how the cellular phone was imagined in different ways according to national context. I will return later to consider what the mobile tells us about our culture that has adopted it so readily. I will ask how the mobile cellphone fits with changing social structures, why it has become the focus of new types of crime, and what it can signify when it appears in cultural products such as television programmes and movies. For material components alone do not add up to a working cellphone. Indeed, it was the scarcity of a non-material resource that prompted the idea of the cellular phone in the first place.

Chapter 2
Save the ether

When Lars Magnus Ericsson was driving through the Swedish countryside, he still had to stop his car and wire his car-bound telephone to the overhead lines. If he had pressed his foot on the accelerator, the wire would have whipped out, wrecking the apparatus. It was not a mobile phone in our current sense of the word. Until the last decades of the 20th century, most telephones were like this: to use them, you had to stand still, because you were physically connected by inelastic copper wire to the national system. A few privileged people – members of the armed forces, engineers, ship captains – could command the use of a true wireless phone, connecting to the land-locked national system through radio. The reason it was a privilege was because the radio telephone had to fight for a share of a scarce resource: a place on the radio spectrum.

The first radio transmissions were profligate beasts. Take Marconi's again. Such radio waves generated by a spark would crackle across many frequencies on the spectrum, interfering with and swamping other attempts at communication. This problem meant that early radio users had a choice: either find some way of regulating use so that interference was limited, or take a chance with a chaotic Babel of cross-talk. The route to regulation was taken. (Although not in all parts of the world: for many decades Italian radio was the liveliest in the world.) But even when radio circuits became more tuneable, so that radio transmissions could be targeted within smaller bands of frequencies, there was never enough spectrum to go around.

Governments seized the right to regulate the radio spectrum. In the United States the authority was the Federal Communications Commission (FCC), in Britain it was the Home Office and the Post Office, and so on. But since radio waves are no respecters of national boundaries, national governments had to concede that international regulation had to take place too. Here there was a precedent. In the mid-19th century, the question of how to organise global telegraph communications had prompted the creation of one of the first truly international organisations: the International Telegraph Union (ITU), with headquarters in the neutral Swiss city of Geneva. With the arrival of the fixed-wire telephone in the late 19th century it made sense to extend the ITU's powers over the new technology. Likewise, the ITU was on hand to provide an organisational structure to regulate international frequency allocation for radio. Every few years, giant international conferences would decide, given existing and predicted use of radio, which services should be allocated a small slice of rare spectrum. These highly technical meetings reflected the world as we know it: engineers and bureaucrats sought to balance demands for new lifestyle electronics, such as music radio stations, with the commercial necessities of reliable navigation aids, and with the conflicting military imperatives of the technological infrastructures of World War and Cold War armed forces.

In these decades, a phone that worked by radio was a simple enough proposition, but was impossible to imagine as a truly everyday and popular device, since there was no way to squeeze its demands into an overcrowded spectrum already dominated by the powerful commercial and military interests of the 20th century. Each radio would have to work on a separate frequency from its neighbours, otherwise calls would be interfered with, confused, or, worse, eavesdropped. So the radio telephone was restricted to a privileged handful.

But in 1947, engineers at Bell Laboratories in the United States proposed a radically new means of imagining mobile radio. It was already shaping up to be a vintage year for Bell Labs. Pump-primed by massive expenditure to develop electronics during the Second World War, the peacetime years saw a string of lucrative discoveries. William Shockley, John Bardeen and Walter Brattain had devised the transistor, the electronic component announced to the public in 1948 that would sweep away bulky valves and lead to the revolutionary lightweight electronics of the second half of the 20th century. Down a few corridors from the transistor pioneers, D.H. Ring, assisted by W.R. Young, put pen to paper, and the result was a description of the 'cellular' idea. It was a means of saving spectrum.

Chapter 3

The cellular idea

Ring had written down the principles on which your mobile phone works. Imagine a map of a city and imagine a clear plastic sheet, ruled with a grid of hexagons, placed over it. Now, imagine a car, equipped with a radio telephone, driving through the streets of New York City, passing from hexagon to hexagon (see diagram below).

Ring's idea was as follows. If each hexagon, too, had a radio transmitter and receiver, then the radio telephone in the car could correspond with this 'Base Station'. The trick, said Ring, was to allocate, say, seven

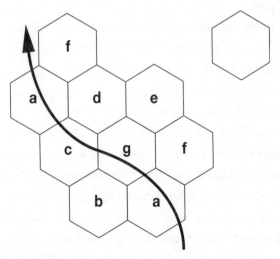

Reusing frequencies creates a pattern of 'cells'.

frequencies to a pattern of seven hexagons ('a'–'g'), and to repeat this pattern across the map. The driver would start by speaking on frequency 'a' in the first hexagon, then with 'g', then 'c', and then back to 'a' again. Now if the first and last hexagon were far enough apart so that the two did not interfere with each other (and this was possible, especially if low-powered transmissions were used), then a radio conversation could carry on without interference, so long as no one else was in that small hexagon, on that frequency, at the same time. If the repeated pattern of hexagons spread over the entire map, then the whole city could be covered. Suddenly, if the hexagons were made small enough, many more mobile telephones could be crammed into a busy city, and only a few of those scarce frequencies would be needed.

But notice some of the consequences of Ring's idea, if it was to be put into practice. The driver of the car certainly did not want to have to know when the car was passing from one hexagon to the next. Indeed, when you are walking down the high street now, you are passing from cell to cell, yet this 'call handoff' or 'call handover' between cells goes mercifully unnoticed. To conceal the handover, the system needed to be able to spot when the mobile was leaving one hexagon, to find the next hexagon, and to hand over the call. In turn, this meant that the individual phone had to be identifiable; so, somewhere in the system, a database needed to be at hand containing information about where the phone was, where it was heading, and who was using it (so the call could be billed). That database had to be fast, so it had to be automatic and electronic.

Furthermore, say that the driver of the car was in fact talking to his best friend, who was in a phone box in Philadelphia 100 miles away. As the car left and entered each new invisible hexagon, the conversation would be seamlessly pieced together at some central Mobile Switching Centre (MSC), the heart of any cellular network, and then fed into the system of old landlines and exchanges, so that finally the distant driver's voice could be heard in the phone box in Philly. For the cellular idea to work, a whole *fixed* infrastructure needed to be in place: base stations, a mobile switching centre constantly interrogating a database of personal and geographical information, and connections to the old Public Switched Telephone Network (PSTN). Just as the pocket watch required fixed institutions of agreed protocols and time standards in order that time could be told on the move, so a massive fixed infrastructure of wires, switches and agreements needed to be in place for mobile conversation. Mobility, strangely, depends on fixtures.

Ring had described the cellular idea in 1947, but it went unpublished and, for nearly two decades, gathered dust. Why was this? Partly, the reasons were technical: cellular phones would work best with higher frequencies, where transmissions could be limited to smaller hexagons and where the spectrum was relatively uncongested. But radio engineers had only slowly gathered the expertise to work at higher and higher frequencies through the 20th century. To find the radio stations found on 1920s bakelite sets, the listener had tuned to up to a few hundreds of thousands of Hertz (Hz, or cycles of radio waves per second, a measure of frequency). During the Second World War, the demand for better radios and the development of new technologies such as radar had ramped the usable frequencies up to many millions of Hz (Megahertz or MHz). The first cellular phones worked in the 800–900 MHz band. In 1947, the techniques to handle such frequencies were cutting-edge science. Furthermore, each time the mobile passed from cell to cell, different transmitting and receiving frequencies were used. In turn this meant that the mobile had to contain a frequency synthesiser, an expensive piece of circuitry to produce the different frequencies: when it was first developed for the military in the 1960s it alone would cost as much as a very good car. Partly, too, the technology of switching in the 1940s was incapable of handling the quick call handovers that the cellular idea demanded. The typical switch of the day was that found in the telephone exchange: the clattering, electromechanical relay that was too slow to implement Ring's cellular concept.

But the answer is never simply technological, for technologies reflect the political and social world in which they are conjured up. Turning Ring's blueprint into a working cellular phone system in the 1940s would have demanded the latest electronic techniques and a crash course to develop new ones, but more importantly the personal mobile phone fits in with social values which dominate now but did not then. The social world of the mid-20th century was hierarchical, paternalistic and even, in large swathes of the globe, totalitarian. The governing model informing both government and business was large-scale and top-down. The telecommunications of the mid-20th century reflected – and in turn bolstered – this pattern.

I was brought up in Hitchin, a small town to the north of London. In the 1960s and 1970s it was a nice place to be. We lived in a house on the outskirts, with fields out the back and the rest of the estate out the front. We had installed, like everyone else, a squat, heavy, black dial-up telephone by our front door. (It was a Type 300 telephone. The design had been copied from an Ericsson phone: the Prince of Wales had seen it at the Stockholm Exhibition in October 1932, and liked it.) The telephone was not our possession, but was the property of a government department, the General Post Office. Only a GPO engineer could take it to pieces if it was broken. (If I had smashed it up, like my Siemens S8, I would have been in trouble. Arguably, it would have been treason, a crime against the state.) If the telephone could not be fixed then only the GPO could supply a replacement. It would be identical – in ironic memory of the long-dead Henry Ford, it seemed as if the Post Office still insisted that you could have any colour, as long as it was black. (Other exotic colours – such as ivory – were supposedly available, but I think they were mythical.) The replacement might take a while to arrive. There was a waiting list. A long one.

Nothing better expresses the difference between then and now than this sentiment: we did not mind. We were happy with the squat, black, unreliable telephone. The personal cellular phone would have been almost impossible to imagine in such circumstances.

The black telephone of my childhood. The Type 300 was introduced in 1938 by the General Post Office and lasted for decades. (BT Archives)

To say that it did not happen because the technology was not there misses an important part of the story. Technology only becomes 'there' when it fits the wider world. Sometime between my early days, growing up in Hitchin, and now, with the pieces of my old mobile phone in front of me, a revolution has happened. A revolution has swept away the GPO monopoly over telephony, and that revolution in turn was part of something bigger, a global sea change in both politics and technology. Cellular phones are part of this story, and not merely as flotsam but also part of the tidal wave of change itself.

All countries had monopolies providing landline telephones before the 1980s. Even in the United States, where freedom of commerce and industry was prized, the complex Bell System conglomerate, of which the American Telephone and Telegraph Company (AT&T) and Bell Laboratories were part, acted like a monopoly. (To give a sense of scale and power, before it was broken up in the 1980s, it was *many* times the size of today's Microsoft.) So a common factor in the following stories of how the cellular phone was built in different countries was a context initially dominated by monopolistic, often government-owned, hierarchical telecommunications organisations. Sometimes, these PTTs (Post, Telephone and Telegraph administrations), as they are called, were part of the solution; more often they were the problem.

However, the national stories of how cellular phone systems were assembled are very different. By taking apart my old mobile phone, I've found that many material and non-material components are crucial to its working: not only batteries, aerials, chips and LCDs, but also base stations, databases, telephone wire networks, spectrum space and ideas. Bits have come from Finland, Chile, the United States, Germany, Taiwan, Australia and the Congo. But when such components were originally reassembled in the past, to make cellular phone systems in different parts of the world, the resulting systems were very different. The makers of the mobile phone made history, but not in worlds of their own making.

Part two
Different countries, different paths to mobility

Chapter 4
Born in the USA

Our engineers & inventors have harnessed the forces of the earth and skies and the mysteries of nature to make our lives pleasant, swift, safe, and fascinating beyond any previous age. We fly faster, higher, and farther than the birds. On steel rails we rush safely, behind giant horses of metal and fire. Ships large as palaces thrum across our seas. Our roads are alive with self-propelling conveyances so complex the most powerful prince could not have owned one a generation ago; yet in our day there is hardly a man so poor he cannot afford this form of personal mobility.

This message, buried in a time capsule underneath the New York World's Fair, 1939, was addressed to the human race of the year AD 6939. Putting aside the question of whether the message's expectation that the people of the seventh millennium would understand 'princes', 'palaces' and 'horses' was realistic, the confident, progressive tone rings clear. But the confidence was forced. Not yet a decade of depression after the great Wall Street crash, and with Roosevelt's New Deal – and indeed Hitler's Germany and Stalin's Russia – demonstrating what state action could achieve, the corporations of America were keen to head off criticism, if not revolution, with a spectacular display of how big business was a friend of democracy. Unlike previous fairs, where comparisons between different countries' cultures had been invited, the 1939 New York World's Fair presented the vision of the big

The New York World's Fair, 1939, promised a bright future – so long as big business was left unfettered.

corporations – General Motors, Ford, Chrysler, AT&T, General Electric, Consolidated Edison – and the vision was not of the painful past, but of a bright future.

In the 'Democracity' of the central Perisphere, Edison's 'City of Light', and particularly in General Motors' exhibitions 'City of Tomorrow' and 'Futurama', visitors were given an eagle's eye view of the clean, prosperous, technologically driven near-future. From the corporate laboratories would spring miracles in our lifetimes. There was a cure for cancer, while ladies of 75 had 'perfect skin' and buildings were made of plastic. To the 45 million visitors to the World's Fair, the future was promised to offer personal mobility, not only in the form of cheap $200 cars, but also through phones without cords.

When you left the Futurama display you were given a badge. On it was a simple statement: 'I have seen the future.' This was the message of the New York World's Fair in a nutshell: while many of the technologies would have been familiar to voracious readers of *Astounding, Amazing,* and other pulp sci-fi magazines, the Fair made the technological future look imminent. The message of big business was: stick with us through hard times and very soon this future will be yours. So while handheld personal mobile radios had been a fixture in comics and fantasy literature – Dick Tracy, for example, had a wrist-radio, a watch that could talk – the inclusion of radio handsets in the everyday future of the World's Fair made their reality seem, to Americans, just around the corner.

Indeed, heavy car-bound radio communications had been pioneered in the United States in the 1920s. And, unlike Ericsson's experimental system, these were true mobile radios: there was no need to stop the car. Like the fictional Dick Tracy, the first users were trying to stop crime. With the production of fast cars and good smooth roads, criminals were getting harder to catch. The result was an arms race between organised crime and the police, each in turn adopting faster cars, more fearsome weapons and, to coordinate action, speedier communication. The Detroit police department was first to try the experiment in 1921. The patrolman, alerted by a message, would have to stop the car and call in by wire. But in 1928, a fully voice-based mobile radio system was introduced in Detroit. Other forces followed.

During the Second World War, radio manufacturers, having cut their teeth on police radio, turned to consider military applications. One such company had been started by Paul V. Galvin in Chicago in 1928. But the name of the company, Galvin Manufacturing Corporation, was soon superseded by that of its chief product, 'Motorola' radios, a tag that evokes perfectly the intimate historical relationship between car and mobile radio. One-way Motorola police radios were installed in the 1930s, and the first two-way radio was provided to the police of Bowling Green, Kentucky, in 1940. By then Motorola was gearing up for wartime production. The 'Handie-Talkie' two-way radio was developed for the US Army Signal Corps that year, followed two years later by the 'Walkie-Talkie'. This backpack radio, designed by Daniel E. Noble, worked by frequency rather than amplitude modulation, thereby reducing weight and size while improving performance. Motorola's 35-pound Walkie-Talkie made mobile radio communication practical in the jungles of west Pacific islands or the farmland of Normandy.

All these American systems were mobile radios but not mobile telephones: you couldn't use a Walkie-Talkie to speak to someone in a call-box. Part of the reason lay in wartime priorities. The telephone network and radio remained separate until 1945, when the war's end meant that military production dropped off and new commercial projects could be given the green light. But the separation was also enforced by regulation. The Federal Communications Commission had to be persuaded to drop its opposition before mobile radio telephones could be launched. Nevertheless the FCC granted a licence to AT&T and Southwestern Bell to operate the first basic commercial system, called Mobile Telephone Service, in St Louis, Missouri, beginning in 1946. Soon it spread to 24 other cities.

Demand for car telephones was intense. AT&T launched its 'highway service' between New York and

Boston in 1947, but in New York itself there were severe problems. Not only was there a waiting list of 2,000 potential customers, but 730 lucky users competed to speak on just twelve radio channels. For two decades radio telephony could barely squeeze onto the radio spectrum. Ironically, the congestion was partly caused by the spectacular growth of private radio. '"Mobiling" has become one of the leading activities in ham radio,' wrote Charles Caringella, call-sign W6NJV, introducing his *Amateur Radio Mobile Handbook* in 1965. 'This growth is only natural; more time is being spent in the automobile than ever before – commuting to and from work, as well as weekend and vacation trips.'

There was not enough room on the radio spectrum for private radio and mobile telephony. But in Ring's concept of the cellular phone, there existed a way out: by reusing radio frequencies in repeating cells, spectrum space could be saved, and more users fitted in. AT&T lobbied the FCC, without success, for a decade from 1958 to 1968. Then, at the same time as the civil rights and counter-cultural movements reached a crescendo, there was a reversal of official attitudes. Room was earmarked in the higher end of the radio frequency spectrum for an experiment in cellular telephony, and the radio and electronics industry was invited to respond. Only Bell Laboratories, the research wing of AT&T, replied by the deadline of December 1971. By 1974, the FCC had decided exactly which frequencies the experiment should use; by then Bell had cracked on with the design and construction of equipment, trying it out at the Cellular Test Bed, built near Newark, New Jersey.

In March 1977, the FCC authorised Illinois Bell, the AT&T operating company for Chicago, to install the first cellular phone system. Ten base stations created the cells for an area covering 2,100 square miles around the Windy City. At Chicago's Oak Park the central switch controlled the base stations and linked the system to the public telephone network. It went live in December 1978. By 1981 2,000 users – the maximum the system could handle – could each speak live from their car to Oak Park via the cellular network and from there, by the fixed wires, to any telephone in the United States. Other manufacturers followed. In 1980, a Motorola subsidiary, American Radio Telephone Service, began delayed trials in Baltimore and Washington. (Martin Cooper of Motorola had filed cellular patents as early as 1973.) In North Carolina, a small cellular company called Millicom adapted a phone made by the E.F. Johnson firm, producing the first portable cellular phone – for those with strong arms.

These early systems convinced the FCC that cellular radio was practical, and should be deployed systematically across the United States. But the sheer scale of the country created problems that would not be encountered, as we shall see, in Europe. It was clearly beyond

41

the resources of any company – except one – to install the base stations, switching centres and marketing operations throughout the United States, from coast to coast, necessary for a national cellular phone system. The one exception was the giant AT&T, the biggest corporation the world had ever seen. But by the early 1980s AT&T was under severe attack from those who would oppose monopolies, leading eventually to the breakup of the Bell System in 1984. Monolithic corporations were not welcome to apply. Moreover, America is made up of an archipelago of cities, with great stretches of rural land in between. The FCC decided it made sense to grant cellular licences by auction on a city-by-city basis, the so-called Metropolitan Statistical Areas (any county with both a total population of more than 100,000 and at least one town with more than 50,000), a format that invited the Bell regional operators – and, after 1 January 1984, the 'baby Bells', such as Southwestern Bell, Bell Atlantic, Pacific Telesis (Pactel), Ameritech, Bell South, US West and Nynex – to apply independently. In each area, to encourage competition, *two* cellular licences were up for grabs. One would typically go to the local Bell, while the other went to a new company. As part of the deal, AT&T set up a subsidiary, kept at arm's length to satisfy the regulators, called Advanced Mobile Phone Service.

The auction of the 90 largest metropolitan areas provoked hysteria. So many applications had flooded in by the June 1982 deadline that the hasty FCC decided that, after awards for the top 30 cities had been made, the other metropolitan licences would go by lottery. This meant that any chancer could land a ten-year licence, and then hawk it around more competent operators. The story was repeated when licences for the less lucrative remaining metropolitan areas and the Rural Statistical Areas were decided by lottery between 1984 and 1989. The last 180 metropolitan area licences attracted 92,000 applications. The end result was that cellular America was extremely disjointed. Each town or city had a different operator, often very local. New companies such as Chattanooga Cellular Telephone, Fresno Cellular Telephone and Long Beach Cellular Telephone served just their local constituency. Roaming, the ability to use your cellphone in different systems, for example to go from San Francisco to Los Angeles and make calls in both cities, was made extremely difficult.

The disjointed pattern was slowly reversed as the industry consolidated. Firms with licences were bought or merged. By 1992, the largest operator, McCaw, had bought out LIN Broadcasting and 90 other licences, to serve a total population of over 65 million people. Two years later McCaw was bought by AT&T – monopolising forces were creeping back. But by then the FCC's management of the licensing process had created a distinctive national cellular style, a crazy-paving of licences covering the country, and had done so slowly. As Garry A. Garrard concluded in an analysis of this licensing phase, the United States had 'spent four years awarding its cellular licences for major markets alone, and seven years in total which, when combined with the initial delay in authorising any cellular service at all, gave other countries the chance to catch up on, and overtake, the technical lead originally provided to the US by AT&T.'

We will see very soon what other countries were doing to overtake the American lead. But first let's sum up developments so far. Both the concept and first working examples of cellular telephony emerged in the United States, albeit with decades separating the two. The most important factors shaping developments were the existence of the world's greatest electronics-based company, AT&T, *and* at the same time hostility to its monopolistic tendencies. The result was innovation, but in a disjointed form, with many small cellular companies rather than one large one. Nor was there a lot of competition, since different companies were restricted to different cities. But there was one standard, called AMPS, after the AT&T subsidiary that gave it its name. This standard dictated how a 'terminal' (the mobile) would communicate with the base stations. We will soon see other standards that would compete with AMPS in the wider world.

If there was a distinct pattern of cellular infrastructure in the United States, there was also a distinct pattern of American cellphone use. Unlike in Europe or Japan, the owner of a mobile phone was charged for accepting an incoming call. This made owners reluctant to give away their mobile phone number to all and sundry, and had the effect of making mobile phones a device for business or emergencies only, and not for chat. The relatively sophisticated – and expensive – cellphone also had to compete with pagers and beepers, which were already very popular with Americans. When I stepped off the plane at Baltimore airport in the 1990s the effect, for me, was obvious to see: while in London the mobile was ubiquitous as the prime means of keeping in everyday conversational touch with friends and work contacts alike, on the other side of the Atlantic it was hard to spot a mobile being used, and if it was the call was brief and businesslike. This difference forms the foundation of a quite distinct mobile culture in America compared to Europe or Japan.

Finally, there was also a difference emerging in the material design and styling of American phones. For decades, Motorola had led the way in car phones. In early 1984, the company introduced the first hand-portable cellular phone, the Motorola 8000, although since it weighed only slightly less than a pack of sugar, this black brick-sized device was not easy on the elbow. It was hardly an instant commercial success. Four years later, hand-portables only made up 6 per cent of sales. It is to countries such as Finland and Japan that we must turn to find enthusiasm for well-designed and colourful handsets. Garrard offers an explanation for the slow adoption of hand-portables in the United States: 'the high price compared to that for car phones, the fact that early cellular networks were not designed for hand-portable use resulting in variable or poor quality reception; and the fact that the entire US way of life revolved around the automobile'. But the Motorola 8000 was also the fulfilment of another corporate promise made five decades previously: the wireless personal telephone that formed part of the gleaming near-future of the New York World's Fair.

Chapter 5
The Nordic way

American business producing innovative technology is a story of the 'dog bites man' variety. It is not news. But within two decades of the launch of cellular telephony in the United States, European bureaucracy had produced a technology that more than matched it. As surprising stories go in the history of technology, the success of this European system, GSM, was definitely man biting dog.

While so often an obstacle to political or technological harmony, the diversity of Europe also means that a wide range of different initiatives can be supported. The development of cellular phones in Europe, for example, directly stemmed from the unique characteristics of one European area: the Nordic countries of continental Scandinavia (Denmark, Norway and Sweden), plus Finland. Only once the cellular idea had been realised in the north did the system spread, for further political reasons, across Europe.

Take Sweden. In the 17th century, Sweden was a great power, dominating Scandinavia, Poland and the Baltic trade into Russia. (Indeed, the 'Rus' had been an ancient name given by the Slavs to the colonising Swedish population.) Its style of monarchical government was absolutist (like most of Europe), bureaucratic (like much of Europe), and relatively efficient (unlike much of Europe). This feature, due to the integration of experts of all kinds into the state, survived, even as once-great Sweden declined. Even today experts – such as engineers and academics – can move between university and government with far greater ease than in other countries.

Sweden industrialised late, from the 1870s, without forming massive smoke-stack cities like Manchester in England or Lyon in France. The new industries – pulp, paper, ball bearings, matches (Ivar Krueger, the 'match king'), explosives (Nobel), weapons (Bofors), steel, telephones (Ericsson) – employed scientific experts but not many other people, compared to, say, cotton mills. Instead, the unemployed rural population upped sticks and emigrated: a quarter of men, women and children left Sweden, mostly bound for the United States. This had several effects. A sympathetic bond was forged across the Atlantic, which helps explains why American trends, for example of efficiency and corporate research, could be adopted back home. Most importantly, however, the declining population made government and the owners of industry very unwilling to upset the labour force, resulting in a distinctive political style: the building of consensus and the rational arbitration of disputes by experts, social democracy and the welfare state. In this society, unlike England or Germany, technology was not generally regarded as a threat – even to jobs – but as something good for every-one. Technology was an equalising force. This aspect of the consensus between industry and individuals was reflected in a passion for industrial design (think IKEA), opinion polls, and apparently democratising but ulti-mately paternalistic technologies such as radio.

Swedish industrialisation was also marked by its stance towards natural resources, in particular the swathes of forest in the north. This Norrland provided the raw materials for the pulp, match and paper indus-tries, but it was not ransacked. Instead Swedes regarded it as a national resource, and one to be used efficiently in a planned, rational manner. Indeed Sweden had, until very recently, a highly dispersed and sparse popu-lation, encouraging identification with the countryside. An environmental tinge entered into the collection of ideas and attitudes that Swedes regard as their contri-bution to the world. Fed by a nostalgia for great power status, by a 20th-century foreign policy of military neu-trality, and by pride in the harmony created by 'third way' social democratic politics, Swedes have been keen that the world should learn from their example. Since the Second World War, Swedes have, for example, been highly active in the United Nations. We will see that this peculiar internationalism shaped technology too.

Swedes tend to bristle indignantly at the suggestion that they are very similar to Norwegians or Danes. Of course no two individuals are alike. But Scandinavian countries share more cultural and political similarities than differences, however jealously cultivated the latter can be. A shared history helps explain this: the three formed the 15th-century Kalmar Union, which, with different twists of fate, could have been the foundation of a United Kingdom of Scandinavia. In the 19th century, under Oscar I of Sweden, an explicit project of 'Scandinavianism' was pursued, until 1864 when its poverty was revealed as Sweden failed to come to Denmark's aid against a Prussian invasion. Still, many of the characteristics I have ascribed to Sweden can also be applied to Denmark, Norway and to some extent Finland, including social democracy, internationalism and an enthusiasm for technology.

In 1967 the chief engineer at Swedish Telecom Radio, Carl-Gösta Åsdal, suggested that an automated nationwide mobile telephone and paging network should be built, integrated with the landline network. Further studies, supervised at Swedish Telecom's radio laboratories by Ragnar Berglund and Östen Mäkitalo, tested the feasibility of Åsdal's idea. For such an ambitious project, collaborators were needed. In 1969, the Nordic Mobile Telephone (NMT) Group was established, with engineers representing Sweden, Denmark, Norway and Finland, and with a view to developing a cellular phone system. These engineers worked for the state-owned telecommunications businesses, and they therefore carried with them the shared values of

Nordic government, in particular a faith in development by consensus and rational discussion between
experts. (These common values would not have been
found if, say, the gathering had been between English,
French, German and Spanish electrical engineers.) The
NMT standard, defining how parts of the cellular system would interact, was also crucially affected by the
Scandinavian relationship between expert and state:
while the expert was integrated into government, and
therefore retained influence, a close touch with everyday society was preserved. In the spirit of good industrial design and of technology as an equalising force,
customer surveys ensured that the NMT standard
matched people's wants and desires. (But notice how
different the Scandinavian engineers' attitude was to
the General Post Office's, say. No insistence on one
squat black telephone here!)

The political heritage can also be found in
Scandinavian management of the radio spectrum.
Like a virtual Norrland, the spectrum was seen as a
national resource, the use of which should not be
restricted but be planned for efficiency and the greater
good. Already in 1955, for example, the Swedish state-
owned telecoms company, Televerket, had launched
the first mobile telephone system in Europe, less than
a decade after AT&T's equivalent. With a small population spread over a big, forested country, mobile radio
found many customers. With high demand and official

encouragement, there were already 20,000 mobile phone users in Sweden by 1981, proportionately far higher than elsewhere in Europe, when the NMT cellular system was launched. With such a prepared and fertile ground, it is not surprising NMT was a success. By 1986 capacity was full, and a second, higher-frequency system (called NMT 900 to distinguish it from the lower-frequency NMT 450) was introduced.

NMT had demonstrated that it was possible for electrical engineers from different countries to sit down together and write a standard for a trans-national cellular phone system. It was possible to 'roam', to use the same phone passing from Oslo to Helsinki, because the technical details determining how a mobile terminal would communicate with base stations, or how base stations would link to the switching centre, had been hammered out beforehand. (Compare this ease of roaming with the situation across the Atlantic, where the common standard, AMPS, was much looser, and where operations depended to a greater extent on decisions made by each company. It was impossible to roam freely across the privately-owned crazy-paving of American cellular.) Furthermore, once the NMT standard was defined, the telecoms industry, already closely involved in developments, could freely offer products. Mobile phones for the pioneering Nordic cellular system were provided both by home companies (the Danish Dancall and Storno, Swedish Ericsson, Finnish

Mobira – part of Nokia), but also by American and Japanese. The prosperous Scandinavians could afford the price – the equivalent of the small car that was still needed to carry an early mobile phone. However, the contract for the central switching technology, the heart of any cellular system and therefore a strategically important economic decision, went straight to Ericsson.

Chapter 6
Europe before GSM: *La Donna è Mobile, Männer sind nicht!*

By 1987, five years after the launch of the Nordic mobile cellular phones, roughly 2 per cent of the population were subscribers. Cellphones had become a standard tool for truckers, construction workers and maintenance engineers, although a few were being sold for private use, especially for installation in the weekend holiday homes and boats that are a feature of Scandinavian life. This early start would continue into the 21st century.

The public telecoms monopolies in other European countries slowly began to respond to the Nordic experiment and to glimmers of public demand. Several – Spain, Austria, the Netherlands and Belgium – ordered NMT services. Often these services stalled because the exact NMT frequencies were already in use, which meant that new expensive mobile terminals had to be produced. In Spain, for example, Telefónica had ordered the NMT system from Ericsson even before it had been launched in Sweden, but the pricey terminals attracted few customers.

But the chauvinistic telecom monopolies of the 'big' European countries – Germany, France, Italy and Britain – decided to design their own mobile cellular systems. Each was designed to respond to peculiar national demands. In France, the Direction Générale Télécommunications began an ambitious *grand projet*, Radiocom 2000, which was more like a souped-up national dispatch radio system than a cellphone. It was as distinctly Gallic as Minitel, the network of public information terminals that would later act as a barricade against the internet-pronged attack of American mass culture. Contracts for Radiocom 2000 were awarded to another immense state-controlled French combine, the military and aerospace firm Matra. Like many French national projects, when Radiocom 2000 was launched in 1985 it was available only in Paris. Again prices were high and uptake low. However, it is easy to be unfair in retrospect. Mobile analyst Garry Garrard records the telling observation of France Telecom's Philippe Dupuis in 1988: 'If it had not been demonstrated in other countries that mobile communications can become more abundant and cheaper, everyone would be happy.'

Likewise, in Germany, the country in Europe in which wealth and technical proficiency were highest, a new cellular standard was developed by the telecoms monopoly, Deutsche Bundespost, and the dominant electronics manufacturer, Siemens. Netz-C was launched – comprehensively, to 98 per cent of the West German population – as a commercial service in May 1986. A unique feature of early German mobile cellular phones was a personal identification card, the ancestor of the later SIM card. This bonus helped bump the price up, so again, compared to cell-happy Scandinavia, sales were disappointing. Politics shaped the German mobile system in another unique way. In 1990, following the fall of the Berlin Wall and the subsequent reunification of Germany, the appalling state of East German telecommunications had to be confronted if the wider tasks of transforming the economy and rebuilding government were to proceed. The situation needed a demonstration of the power of capitalism. Instead of slowly installing telephone cables, the mobile Netz-C was quickly expanded to cover the East German Länder.

But Netz-C was a poor system. Compared to Italy's, and especially the pioneering Nordic cellphones, German mobile was a failure. We will see soon how an unsatisfied German market would be crucial to launching a new, world-beating mobile system. However, if we had paused in 1991 and surveyed mobile Europe, we would have discovered a pattern that would have confirmed the prejudices of any Eurosceptic. The picture was of a hotchpotch of national systems, each designed to satisfy national interests (often, indeed, the interests of the public telecoms monopolies rather than French, German, Swiss or Spanish customers), and employing ten incompatible standards. Oddly, the spectacle was most pitiable in the most wealthy, technophilic countries, France and Germany. The United Kingdom, insular in its own way, was pursuing a detached Thatcherite experiment. We will see soon what happened there. But first, a bureaucratic miracle.

Chapter 7
GSM: European union

In a chilly Stockholm, in December 1982, engineers and administrators from eleven European countries gathered to inaugurate GSM – at first an acronym that merely described themselves, the Groupe Spécial Mobile, but later, as the Global System for Mobile, something that would record an unlikely feat of world conquest. The lead had come not from the big European powers, but from the Nordic countries, which had in NMT a successful trans-national mobile system ready for expansion, and from the Netherlands, which did not. The delegates were there to consider whether a Europe-wide *digital* cellular phone system was technically, and more importantly politically, possible. Older standards such as NMT had been analogue; good for the transmission of voice, but little else. Going digital created the opportunity to provide new services such as data transmission, but also, more importantly, the chance to make a pan-European political statement. But, despite the example of NMT, surely the odds of overcoming entrenched national telecoms interests were slim?

In February 1987, in the warmer climate of Madeira, the group gathered again to hear the results of proto-type tests. Not only had these tests been passed with flying colours, but the political will had been found to iron out national differences and to choose one, pan-European standard. What had happened?

The 'Europe' built on the ruins of the Second World War was the outcome of two opposing tendencies. For every Jean Monnet or Robert Schumann who dreamed of a United States of Europe, there was a General de Gaulle or Margaret Thatcher who was deeply suspicious. It is noticeable, however, that while the sceptics of Europe bolstered their arguments by appeal to grand, even sentimental *ideas* – of national spirit or self-determination – the federalists' project of building Europe has often progressed by mundane and technocratic appeal to better technological, *material* organisation. So, for example, following the Monnet-authored Schumann Plan of May 1950, the European project was launched with the establishment of the European Coal and Steel Community by the Treaty of Paris in 1951. In six countries (France, Germany, Italy, Belgium, Luxembourg and the Netherlands), the industries that produced the material structure of Europe – the Europe of reinforced concrete, car chassis, dynamos and steel knives – came under the power of a supra-national Higher Authority. In 1957, with the two Treaties of Rome, one community became three, with the addition of a European Economic Community and, in the nuclear field, Euratom – another example of the technological spirit behind European organisation. In 1986, by which time the original six had been joined by the United Kingdom, Denmark, Ireland, Greece, Portugal and Spain, the Single European Act expanded on the earlier target of mere coordination of the EEC, becoming the much more robust aim of making Europe a single market, bigger than any other in the world. The project to build one European cellular phone system, based on the GSM standard, would be a major material means of realising the dream. As the institutions of Europe grew more powerful, so they increasingly became the source of further pushes towards European integration. Of these, the European Commission, created from the smaller staff and management of the three existing communities and headed by a President, was the most important.

Evidence for GSM as part of a dream of 'Europe' can be found by wading a little way into European bureaucracy. In particular a Recommendation issued by the Council of the European Communities on 25 June 1987, called 87/371/EEC, addressed the question of mobile communications directly. The Council is a collection of politicians, but the paperwork it considers is largely generated by the Commission. So 87/371/EEC should be seen as an expression of the interests of the European Commission. The document is written in legalese, but the reasons for favouring a pan-European mobile system shine through. Broadly, they were two-fold.

First, the dream of a single European market would remain just that if a means was not found of reducing national differences and improving communication. When the authors of the document wrote that 'the land-based mobile communications systems currently in use in the Community are largely incompatible and do not allow users on the move in vehicles, boats, trains or on foot throughout the Community, including inland or coastal waters, to reap the benefits of European-wide services and European-wide markets', they were addressing both problems. GSM offered an exceptional moment for reducing difference: the 'change-over to the second generation cellular digital mobile communications system' provided 'a unique opportunity to establish truly pan-European mobile communications'. 'European users on the move', communicating 'efficiently and economically', would be the basis for a single market. Again, 'Europe', an otherwise rather ghostly entity, would be given substance by building material technological systems.

Second, the eurocrats kept a watchful eye on the main economic competitors: Japan and the United States. GSM would not only be an instrument of European unification, but also provide a lead in the cutthroat but strategically important global marketplace for technological goods. 87/371/EEC states explicitly that 'a coordinated policy for the introduction of a pan-European cellular digital mobile radio service will make possible the establishment of a European market in mobile and portable terminals which will be capable of creating, by virtue of its size, the necessary development conditions to enable undertakings established in Community countries to maintain and improve their presence on world markets'.

So the European Commission, the civil service of the European project, had seen in GSM a political tool of immense value: telecommunications – and particularly GSM – would provide the infrastructure of Europe ready to mount a convincing economic challenge to the United States and Japan, and a pan-European telecoms network would encourage organisations to think European. The one drawback was that the European Commission had not been responsible for encouraging GSM from the outset. Instead, the representative of the old, increasingly unfashionable public telecommunications monopolies, the Conference of European Posts and Telecommunications Administrations (CEPT), had been. Nevertheless, with bureaucratic *sang froid*, the Commission claimed credit anyway.

The best illustration of GSM as a politically-charged European project is given by the facility to roam. Just as in NMT, roaming – the ability to use the same terminal under different networks – was prioritised, even though it was expensive, because it demonstrated political unity. The complexity of the technical specifications that allowed a mobile user to drive from Lisbon to Leiden gave GSM a new, unwelcome nickname – the 'Great Software Monster'. Commentators called GSM the 'most complicated system built by man since the Tower of Babel'. But the political intention was in stark opposition: in place of the conflicting chaos of incomprehensible tongues, GSM would stand for unity. In practice, very few users had roamed across Norway, Sweden, Denmark and Finland with the NMT system. Most users had stayed within a few miles of Oslo, Stockholm, Copenhagen or Helsinki. Again, with GSM, for many years roaming was an expensive political luxury – the telecommunications equivalent of the agricultural subsidies that were grudgingly paid to keep European peace. Ironically, when roaming did take off in the very late 1990s, it was enthusiastically embraced by partying twentysomethings in Ibiza as much as by European business executives. The reason can be found in a capacity that had been buried in the GSM specifications as little more than an afterthought: the Short Message Service. The phenomenon of text messaging will be examined later.

GSM was intended to be launched on 1 July 1991. In fact it wasn't really ready. But the bureaucratic miracle had to be witnessed, and a few symbolic conversations were organised: for example Harri Holkeri, the Governor of the Bank of Finland, phoned the mayor of Helsinki, and discussed the price of Baltic herring. But such fishy anecdotes aside, commercial GSM services did not really start until the following year. Eight countries – Germany, Denmark, Finland, France, the UK, Sweden, Portugal and Italy – began in 1992, and were soon joined by others, so that by 1995 European coverage was nearly complete. Indeed, several countries had more than one GSM operator. Remarkably, GSM then began to be adopted outside of Europe. By 1996, GSM phones could be found in 103 diverse countries, from

The GSM standard for digital cellular phones was a worldwide success. This shop in Dalston, London, specialises in GSM phones for African networks.

Australia to Russia, from South Africa to Azerbaijan, and even in the United States.

European bureaucracy had scored an undoubted commercial hit. There were many factors behind the remarkable uptake of GSM. A feature of the history of standards is that success tends to create its own momentum. So, to take two notorious examples, the QWERTY keyboard and the VHS video system were both 'chosen' not because they were technically superior to their rivals (they weren't), but because everyone else was already using them. GSM was not the only digital cellular standard on offer in the 1990s, but since a significant number of countries had already adopted it – albeit for political reasons – then it was a safer choice. The equipment was ready to buy, and experience showed that it worked well enough. But this dynamic is not by itself enough to explain GSM. The European digital standard benefited, bizarrely, from the chaos of what went before. Many different national systems at least produced a variety of technical possibilities from which to pick and choose. NMT provided a basic template onto which extra features – such as SIM cards, from Germany's Netz-C – could be grafted. Furthermore, the relative failure of the first cellular phones in countries such as France, and especially Germany, created a pent-up demand that GSM could meet. In the United States, where customers were satisfied with the analogue standard, there was little demand for digital until spectrum space began to run out. Paradoxically, America lost the lead because its first generation of cellular phones was too successful.

But more important still was how GSM satisfied both European customers and manufacturers. Recall that when I smashed up my old GSM phone I found that it consisted of many components. In the early 1990s technical trends, especially miniaturisation, led to a qualitative change in mobile terminal design. Suddenly, mobile phones became small and light enough to routinely carry around. ('Handportables' had existed before, but they were not cheap and certainly not easily manageable.) There was a leap from car phone to hand phone. Part and parcel of the same process, the new designs attracted new customers, and the mobile became less a business tool and much more an everyday object. The shift marks the start of the mobile as an object of mass culture and individual necessity. Three manufacturers soon dominated the mobile market. Spurred by competition between each other, Nokia of Finland, Ericsson of Sweden and the American Motorola designed and marketed ever smaller and cheaper phones.

These companies, and European firms that supplied other parts of the GSM cellular networks, also benefited from the fortunate outcome of a patents battle. Aware that such a complex system as GSM might rely on patents that may only be discovered, ruinously, at a later date, European planners of GSM insisted in 1988 that manufacturers indemnify themselves against the risk. Although the demand was later watered down, many manufacturers, including the mighty Far East conglomerates and most US firms, decided to sit GSM out. This obscure legal controversy had the effect of reducing competition and massively strengthening the firms amenable to agreement: Nokia, Ericsson and Motorola. (Motorola had European factories.) Select manufacturing interests were therefore happy with GSM. These factors were as important in pushing GSM forward as pan-European political motives had been for starting the ball rolling.

Chapter 8
Digital America divided

Only when too many American analogue cellphone users were cramming into the available spectrum space did attention seriously turn to launching a digital standard that could compete with GSM. In the spirit of free market competition, a number of alternative approaches were proposed.

Imagine you have 30 people at a great party, all on the same cellphone network and all calling their friends. If they all tried to use the same frequency at once, then no one would be able to talk. This was the problem confronted by the telecommunications engineers at the very end of the 1980s – except that the party could be the size of Chicago or New York. There are several solutions. Each party-goer could be allocated a tiny sliver of frequency of which they had sole use. This solution, called Frequency Division Multiple Access (FDMA), only works so far as you can slice up the scarce frequency bands thin enough without affecting operation. In practice, you soon run into problems. So another solution is to not broadcast all the conversation. For example, take a second of transmission. By devoting the first thirtieth of a second to the conversation from the first party-goer, the second thirtieth of a second to that of the second party-goer, and so on, a small part of every phone call from the party is trans-mitted. A trick is played on each listener – they don't notice the gaps! This technique, easy by digital meth-ods, is called Time Division Multiple Access (TDMA), since you are dividing up the transmission into differ-ent time slots and letting lots of callers access it.

A third possibility is much harder to explain. It is best attacked from a different direction. Imagine some-one at the party wanted to have a secret, illicit conver-sation. The caller could borrow a technique from the shadowy world of codebreaking. By mixing up the information that coded the secret conversation with a seemingly random signal, and transmitting the mix-ture together, then the caller could ensure that even when this was mixed in with other messages using the time division method, the receiver at the other end – if they knew what the random signal was – could unpick the original message. What this method – called Code Division Multiple Access (CDMA) loses in simplicity, it gains in privacy. The invention of CDMA had its roots in San Diego County, where there existed a fruitful inter-play between military contractors and top electronic research centres, such as the University of California at San Diego (UCSD). Irwin Jacobs and Andrew J. Viterbi had met at UCSD in the 1960s and launched Linkabit, a military communications company, in 1968. In 1985 they sold out and started again with Qualcomm, through which Jacobs and Viterbi hoped to commercialise another product of the Californian military-industrial complex: CDMA, a concept that had been developed when Linkabit had been asked to develop a satellite modem for the United States Air Force. Qualcomm had the skills to take CDMA from its military setting and to reap the considerable profits of entering the consumer communications market. (There was also a contract from the Hughes aerospace giant to help things along.) Nevertheless Qualcomm needed considerable powers of persuasion, since the decision was taken in 1992, on competition grounds, to allow both TDMA- and CDMA-based standards to proceed despite CDMA coming late to the party. The result was that the United States yet again resembled a patchwork, this time of a variety of incompatible digital standards.

This pattern of division and fracture lasted until 2005 and the launch of Apple's iPhone. For deep reasons – rooted in Apple's culture of end-to-end control of how users experience their electronic products, which will be discussed later – once the iPhone had achieved its remarkable success, American cellphone culture began to look and feel different. Nevertheless, there was still a divergence between the United States and the rest of the world, especially the developing world: while by 2009 over 85 per cent of Americans possessed one, they remain more reluctant to rely on their cellphone, or even to use it routinely, compared to many non-Americans. Some, although very few, reject the compulsion for a life in constant touch. Nevertheless, it is the old, the poor and the less educated who are less likely to have a cellphone at all. The technology, too, best exemplified by the iPhone, reflects divergent values. Writing for the *New York Times* in 2010, Anand Giridharadas neatly summed it up when reflecting on American cellphone culture on his return from a visit to India:

> Not for the first time, America and the rest of the world are moving in different ways. America's innovators, building for an ever-expanding bandwidth network, are spiralling toward fancier, costlier, more network-hungry and status-giving devices; meanwhile, their counterparts in developing nations are innovating to find ever more uses for cheap, basic cellphones.

Chapter 9
Mob rule: competition and class in the UK

In the early 1990s, on trains across Britain, something exceedingly disturbing was happening. People were talking. Loudly. The anger, generated amongst unwilling eavesdroppers and aimed at the mobile owners cheerfully declaring that they were 'on the train', was a sure indicator that an invisible social boundary had been transgressed. In the early 19th century, the stagecoach had been alive with gossip and chatter, as the novels of Jane Austen or the essays of William Hazlitt record. With the arrival of the steam locomotive, however, the talk stopped. Partly the smooth, speedy – almost unworldly – motion of the railway carriage on iron tracks was more conducive to contemplation of the landscape outside the window than to discussion with fellow passengers. Trains transported the body *and* the mind. But a more severe problem lay with those passengers themselves. *Who* were they? Railways were symbols of an industrial age, and in the sprawling industrial city people became increasingly anonymous. Although the division of carriages into different classes – first, second and third – gave some clues, it remained extremely awkward to strike up a conversation on a 'suitable' note. Rather than

'I'm on Westminster Bridge.' Telecom Pearl in 1986.
(BT Archives)

commit a social gaffe, travellers on trains in Britain chose silence.

Delicate issues of class had created social protocols of communication – rules governing when to speak and what could be said, rules that may never have been written down but were more powerful for all that. Against a century and a half of mutually-sanctioned quiet ran a device created by a new set of protocols. GSM, at heart, was also a set of rules governing communication. But these were hard and explicit, and individuals – not society – accepted them upon the purchase of a cellular phone. It was the individual, not society, that spoke loudly: '*I*'m on the train.' Listeners were annoyed not only because the older tacit rules had been broken, but because their particular complaint against the owner of the mobile phone was of *selfishness*. How dare they disturb everyone else! What makes their conversation so necessary, so important, to justify shattering the collective trance? What makes *them* so special? Indeed, the history of mobile phones in Britain is intimately entwined with social transformations, class transgressions and competition – not only between technical systems but also between the politics of selfish individuality and the social bonds that tie us.

In the summer of 1954, the Marquess of Donnegall was jealous. The Duke of Edinburgh, he had heard, possessed a telephone built into a car. The Marquess's information was correct. The Duke's Lagonda coupé sports car had a radio telephone with which, via an Admiralty frequency and a Pye relay station up on the Hampstead hills in north London, he could speak directly to Buckingham Palace. He enjoyed this perk of the job, as the breezy *Daily Sketch* told its royal-friendly readership: 'The Duke takes a keen delight in making surprise calls to the Queen ... Sometimes he disguises his voice when speaking to Charles and Anne.' (The newspaper also hinted at fears concerning the combination of royalty and speeding cars: 'He is a skilful driver but some concern was felt that he should use so fast a model,' while adding the reassuring statistic that the Duke, as Lieutenant Philip Mountbatten, held the unofficial record among his fellow naval officers for the 98-mile run from Bath to London. In his 12hp MG he had covered the distance in one hour and 40 minutes.)

The Marquess of Donegall asked the Post Office whether such a radio telephone could connect to the public network, and if so, whether he could also have a set. The Post Office's reply is revealing in what it tells us about British mobile radio telephones in the mid-1950s:

> You probably know it is possible for passengers on certain ships to make radio-telephone calls to the United Kingdom public telephone system; and shipping in the Thames, if provided with the appropriate equipment, can also call on-shore telephones. There is, however, no arrangement for private persons to fit radio in their own vessels or vehicles for communication with the public telephone network. The prospect of starting such a service in the United Kingdom in the present state of technical knowledge in the radio field are nil. There is no room in the radio frequency spectrum.

Actually the Duke's car phone could have been connected to the public telephone system – experiments had proven this capability, but the Duke had baulked at the cost. While it was 'Post Office policy to refuse the connection of radio calls from privately operated services (which include those established by police, fire, and public utility organisations) because of the difficulty of maintaining the necessary standard of transmission', an exception could be made for a person of appropriate social standing. In the mid-1950s, if you were the husband of the Queen then you could have a mobile telephone connection to the public telephone network. But if you were a mere Marquess you could go whistle.

But, aside from the finer points of social rank, the Post Office's reply also illustrates the typical sentiments of a public telecoms monopoly. It looked inward, rather than outward. The Post Office was more concerned to preserve the integrity of the network than to be led by – or even concede to – customer demands. Early users of mobile radio in Britain included the travelling car-repair services (such as the Automobile Association or the Royal Automobile Club), taxi firms (particularly in the capital, such as RadioCabs (London) Ltd), and industrial companies whose facilities were dispersed widely and in far-flung places (such as Esso Petroleum Ltd). Even banded together to form the Mobile Radio Users' Association, they were powerless against the might of the public telecoms monopoly. In 1954, for example, the mobile users were kicked off their frequencies because the government wanted to create room in the spectrum for the commercial rival to the BBC, ITV. The Post Office consistently (with few exceptions, not least the case of the Queen's husband) refused to consider connection of mobile radio to the telephone network. By 1968, when, against two decades of Post Office disinterest, there were 6,100 private mobile systems licensed, a total of about 74,000 stations altogether, and a growth rate of 17 per cent per year, the official attitude was still that the integrity of the telephone network was paramount and any interconnection of the noisy, anarchic mobile radio to the state-owned system could barely be countenanced: 'The policy of refusing connection of private mobile systems to the public network has lasted nearly 20 years. From the [official] point of view the argument for maintaining this refusal is as strong as ever.'

Actually, by the mid-1960s the Post Office had begun, reluctantly, to change its policy on interconnection. An experimental South Lancashire Radiophone Service had begun around Manchester, Liverpool and Preston in 1959. In 1965, an extremely exclusive and expensive service, called System 1, had been launched in the well-to-do Pimlico area of West London. It was used by the chauffeurs of diplomats and company chairmen. The radio set cost £350, the service cost over seven pounds a quarter year, and calls cost one shilling and threepence for three minutes. Two years later,

The London Radiophone Service was launched by a call from the Postmaster General (the government minister responsible for post and telecommunications) to the TV presenter Richard Dimbleby. The service was not a cellular system but allowed callers to connect to the public telephone network via an operator at the Tate Gallery Telephone Exchange and base stations at Kelvedon Hatch (near Brentwood, north-east of London), Bedmond (near King's Langley, north of London) and Beulah Hill (in Croydon, South London). (BT Archives)

the emergency services were allowed to connect to the telephone network. Indeed, the mobile telephone had begun to trickle down the British class system. On the eve of the introduction of cellular phones, a privileged 14,000 used the later (non-cellular) System 4 mobile telephones. The TeKaDe terminal alone cost £3,000 and the annual subscription was a quarter of this sum again. Car ownership is a good indicator of status, and most users of System 4 drove a Rolls-Royce, BMW, Mercedes or Range Rover.

The later story of mobile phones in Britain only makes sense in the context of the political

A class act all round. Woman in a Rolls-Royce Corniche, with radiophone, 1975. (BT Archives)

transformations instigated by Margaret Thatcher, who was elected as prime minister in 1979. The early years of her Conservative administration were decidedly shaky. She had ditched the consensus politics of her forebears in favour of rolling back the state and a brutal regime of monetarist economics. Unemployment shot up. Riots flared in London, Liverpool and other cities. Her drastic experiment on Britain would almost certainly have ended in defeat if it were not for two factors. First, the unexpected war with Argentina over the Falkland Islands rekindled nationalist emotions that few, in particular the ineffectual Labour party in opposition, suspected still existed. Second, Thatcherism made an even stronger appeal to the purse than to the national-ist heart. In rolling back the state, Margaret Thatcher realised that the privatisation of state-owned resources released money, in the form of stocks and shares, that could be given to the voter. A virtuous cycle of greed was set up, in which industry was liberalised and individuals enriched. It was to be an experiment in class and competition.

Telecommunications – in other words the Post Office – provided an ideal test case. Two years after Thatcher's rise to power, the Telecommunications Act of 1981 was passed. The phone business side of the Post Office was stripped away and renamed British Telecom, and a competitor was authorised: Mercury Communications Ltd, a new face to the old imperial firm of Cable & Wireless. While for the moment BT would remain a public corporation, like the old Post Office, a second Act in 1984 privatised it. The theory was that with the new exposure to the market, British Telecom would be forced to respond to customer needs. The era of the squat black telephone, with one size fitting all, was over.

But Mercury, competing with British Telecom on the fixed-network telephone service, would provide a poor demonstration of the powers of market capitalism. The leviathan British Telecom retained a *de facto* monopoly over the landlines. Instead mobile telephony became politically hot: since cellular systems would be introduced from scratch, then British Telecom would be on a level playing field with competitors.

In 1982, the government announced that two

analogue cellular licences were up for grabs. As an extra hobble, British Telecom was told that a submission would not be welcome directly, only in the form of a joint bid in a minority partnership. Intensely annoyed, but unwilling to yield a promising area of telecoms business, British Telecom combined with the private security firm Securicor. (Securicor was doing very well in the Thatcher years, and therefore had the money to risk, but it was also an experienced user of private mobile radio.) This service, operating under the name 'Cellnet', was guaranteed the first licence. The second licence was won in December 1982 by a consortium led by the unusually market-oriented defence electronics firm Racal, and inspired by Gerald Whent, then chairman of Racal's Radio Group. Other partners included Millicom, which had operated cellular phone systems in the United States – including a test-rig around Rayleigh-Durham, North Carolina as early as 1981. Racal's group decided to trade under the name 'Vodafone'.

A committee, the Joint Radiotelephone Interfaces Group, on which all the governmental and business parties were represented, decided which cellular standard to adopt. The Nordic NMT was rejected because it would not provide enough capacity for central London. Other possibilities were ruled out for being either proprietary or unproven. In common with much policy during the Thatcherite 1980s, eyes turned to America for inspiration. The American AMPS was a proven standard and would have been ideal, but it operated at frequencies already occupied in the UK. So a tweaked standard based on AMPS, rather grandly called Total Access Communications System, or TACS, was quickly agreed.

The first cellular phone call in the United Kingdom over the new service was made on New Year's Day 1985, fittingly from St Katherine's Dock in the City of London to Vodafone's headquarters in Newbury, 50 miles west in the Berkshire countryside. (One of the callers was the comedian Ernie Wise.) Cellnet launched in the same month. However, neither Cellnet nor Vodafone was an instant success. In a final desperate, and successful, appeal to market forces, a further layer of competition was introduced. Service providers, small entrepreneurial firms, took over the task of selling cellular phones to the public. With the instincts of a barrow-boy trader, these easy-come, easy-go firms aggressively pushed mobile phones to punters. Some private fortunes were made. Many providers would later be swallowed up into more respectable groupings, such as Carphone Warehouse and Hutchison Telecommunications. But by then the cellular phone provided a growing business for Vodafone and Cellnet, and symbolised the 1980s ethos of competition. The brick-like cellphone clasped to the ear became part of the conspicuous consumption exhibited by the City high-flyer – in cliché, the trappings of the yuppie.

A young urban professional uses a Telecom Coral cellular phone as a driverless train arrives on the Docklands Light Railway. The chunky mobile phone and the development of the Canary Wharf site in East London were entwined symbols of the Thatcher years. (BT Archives)

The marketing of mobile phones in the 1980s was aimed squarely at businessmen and women. The following exhortation from 1986, from the instructions issued to the sales force of British Telecom Mobile Cellphones, is especially revealing:

Turning Idle Time into Productive Time

When you're away from your office and your phone, you're effectively out of touch with your business. You can't be contacted. Nor can you easily make contact yourself. Take a mobile telephone – a Cellphone – with you and you get a double benefit. You're totally in touch, ready to take instant advantage of business opportunities when and where they occur. And you can make maximum effective use of 'dead time' – time spent travelling – turning it into genuine productive hours.

A society in constant touch was partly created by this economic rationale of squeezing in ever greater quantities of productive work. Resurrecting 'dead time' – the phrase could equally well refer to time spent with family or at leisure – and reclaiming it in the service of capital.

In 1989, competition for three more licences was announced. While it was hoped that these services, called Personal Communications Networks (PCN), would be highly distinctive – more downmarket than

84

The end of 'dead time'. A businessman is still at work while taking a London black cab, 1985. (BT Archives)

TACS, cheaper than the GSM under way across the English Channel – in fact, they were to all intents and purposes indistinguishable. (The PCN proposals were loathed by the continental Europeans. It was seen by the Germans and French, in particular, as an affront to the spirit of European cooperation represented by GSM. The Eurosceptic Margaret Thatcher was happy with this divergence.) Only eight PCN bids were received. It was by sheer chance that the bidding for the untried PCN and the potentially lucrative German GSM licences coincided, and many international firms understandably concentrated on the surer bet.

One licence was already promised to Mercury (in fact a grouping of Cable & Wireless, Motorola and the Spanish telecoms monopoly Telefónica), so that it could continue in its efforts to compete with British Telecom. Mercury's service was called 'One 2 One'. The other two winners were consortia called Microtel and Unitel. The losing consortia included some of the big names of British electronics, such as GEC, Ferranti (in the form of a spin-off, Ferranti Creditphone) and an innovative private telecoms company based in Hull, Kingston Communications. In 1991, a bewildering sequence of changes in consortia ownership led to just two groupings: a merged Mercury and Unitel, owned by Cable & Wireless and US West and offering One 2 One; and Microtel, owned by Hutchison and the defence firm British Aerospace (BAe), soon to rename itself 'Orange'.

Mercury One 2 One launched in London in September 1993. So when Orange followed in April 1994, four different cellular phone services were available to customers in the United Kingdom. While the two older analogue TACS services (Vodafone and Cellnet) and the two newer digital PCN systems (Orange and One 2 One) were technically distinct, the service was very similar and the consumer could make little technical distinction between the four. (Indeed, all were digital after the mid-1990s.) The result was intense competition based on billing packages and sharp advertising. A revolutionary shift came with the offering of 'pay as you go' packages, which were simple, required no credit check and were anonymous. One 2 One offered off-the-shelf packages with pre-charged batteries – ideal as a gift or for those daunted by complex tariffs. Vodafone promised 'Pay as you talk'. Many of the deals were stoked by the marketing tactic – reprehensible in the view of many continental Europeans – of selling telephones at very low prices, subsidised by airtime revenues.

But marketing was not driven only by price. While all the network operators invested heavily in advertising, it was Orange that made the early running, building a trusted brand with a catchy slogan – 'the future's bright, the future's Orange' – and breezy, amiable ads. Orange also boasted per-second billing and – most important for the fashion-conscious urbanite of the early 1990s – sleeker Nokia phones. The other operators took time – and money – to catch up. Star celebrities helped sell phones. So, in 1996, the supermodel Kate Moss was wishing for a 'one to one' with the young Sun Sessions-era Elvis Presley. (Orange had commissioned extensive public opinion research from the Mori polling organisation, and had discovered that Elvis was 'one of the most popular famous people'.) Meanwhile, all four networks piled on new clients. By the summer of 1994, One 2 One had connected its 100,000th customer; the millionth signed up in January 1998, the 5 millionth in April 2000, and the 9 millionth by 2001.

Cellnet passed 100,000 in 1988, 1 million in 1994 and 11 million by 2002 (by which time the old BT spin-off had puffed itself up with the groovier name 'O$_2$').

Once the expensive business of rolling out the cellular network infrastructures – the base stations, switches, microwave links and so on – had been completed, the revenue from customers rolled in. The cellphone companies soon became industrial giants. In August 1999 One 2 One, less than a decade old, was bought by Deutsche Telekom, becoming part of the worldwide T-Mobile roster of networks, for £8.4 billion. ('T-Mobile' was a contraction of DeTeMobile, the name that had been given to Deutsche Telekom's mobile activities in 1993.) Anglo-German commercial interchange continued the following year when Vodafone, less than two decades old and already combined with the American AirTouch Communications, acquired the German technology and media giant Mannesmann, making it one of the largest companies in Europe and one of the top ten companies, measured by market capitalisation, in the world. As part of the same deal, Mannesmann sold Orange (which it had snapped up in 1999) to France Telecom. By 2001 Vodafone had become an international player, with over 80 million customers across the world. Furthermore, Vodafone symbolised another significant shift: for anyone following British industry for much of the 20th century, the idea of a British firm buying a major German technology-based company would have seemed laughable. But Vodafone's share price, riding a wave of tech-stock enthusiasm, gave it immense purchasing power. However, as we shall see, the market can go down as well as up.

Competition between One 2 One, Orange, Cellnet and Vodafone had brought prices down and made the mobile phone an everyday object. No longer was it a status symbol – signifying privilege in the 1950s or wealth in the 1980s – but instead the universal accompaniment of young and old alike (although particularly the young). As the mobile trickled down the social scale, it became a great leveller: granting the power of mobile communication and organisation to the shifting, roaming crowds. A population in constant touch.

Indeed there was an ironic twist in the levelling powers of the mobile phone in 1992. In August, the tabloid *Sun* newspaper devoted ten pages to a taped conversation between Diana, Princess of Wales and James Gilbey, revealing the two to be lovers. The royal marriage, already strained, was in tatters. In December, the same month that Windsor castle burned, the prime minister, John Major, announced the divorce of the Prince and Princess of Wales to the House of Commons. The source of the *Sun*'s revelations, the 'Squidgy' tapes, so called after Gilbey's pet name for Diana, had been recorded by eavesdropping on a conversation over the princess's mobile phone. It was a sign of a major scandal to come.

Chapter 10
Decommunisation = capitalist power + cellularisation

In 1920 Vladimir Ilyich Lenin, leader of revolutionary Russia, surveyed a country ruined by civil war and racked by starvation, and confidently announced that progress was assured through the rapid construction of a technological network. His slogan was pithy: 'Communism is Soviet power plus the electrification of the whole country.' With the fall of communism across central and eastern Europe seven decades later, the nascent liberal democracies were blessed with a small army of economic advisers from the West. Their mantra, heard loudest in Lenin's homeland, called for the unleashing of entrepreneurial activities, the rapid privatisation of state-owned industries, and the opening of markets to foreign companies. Following close behind the economists were Western cellular phone companies.

We have already seen how mobile phone systems were built in former East Germany as a way of providing communication services without relying on obsolescent landlines. However, the rise of capitalist power in Russia did not coincide with the cellularisation of the *whole* country. Not only was it too vast, but socialist ideals such as universal coverage had been abandoned.

Did you join the network? (With apologies to Dmitry Moor)

A condition of gaining a licence, such as those enabling US West and Millicom's services in Leningrad and
Moscow (both from 1991), was that local partners were
involved. In the capital, licences were granted, sometimes in return for cash donations to the local powers-
that-be, for a number of different standards – yet again

the style of technological system was being shaped by political context, in this case the turmoil and confusion of post-Soviet Moscow government. At one stage Bell Canada thought it had won a GSM licence, but then backed out when a surprise demand for $50m was made. Likewise, as Garrard records, it was to the 'consternation of operators that had received NMT450 or GSM licences [that] AMPS licences were announced for Moscow and three cities in the East of the country'. One of the Moscow licences went to a consortium led by the Cold War defence giant Vimpel. In 1996 VimpelCom, founded by Vimpel's Dmitry B. Zimin and American investor Augie K. Fabela II, became the first Russian company listed on the New York Stock Exchange. 'Bee-line', VimpelCom's mobile brand, symbolises a new Russia, in which Western investment has put flesh on old Cold War bones.

Likewise in China, the transition from communism to some form of capitalism was reflected by the spread of mobile base stations. Chinese cellular telephony started in 1988 with a TACS system (i.e. the British standard) in Beijing, Shanghai and Guangdong province. Two companies, China Mobile and China Unicom, offered increasingly popular GSM-based services in the 1990s. By 2002, China had become the number one mobile country: there were 160 million subscribers, overtaking the United States. Measured by percentage of population, of course, the picture looked somewhat different (only just over 12 per cent of Chinese, compared to nearly 38 per cent of Americans, had cellphones). The potential size of the Chinese market was clearly huge.

However, Western companies attracted by the market often met a rocky reception. It took nearly a decade of lobbying, for example, for Qualcomm to gain permission to supply a CDMA network to China Unicom, negotiating past tricky moments in China–US relations, such as the bombing of the Chinese embassy in Belgrade in 1999 and the ramming of an American spyplane by a Chinese jet in 2001. But Qualcomm's determination seems to have paid off, and the company hopes to follow Ericsson's example: the firm's biggest market is in China, not Europe or the United States.

Chapter 11
Japanese garden

It might seem strange to have read almost half a book about the development of a new consumer technology without Japan entering the story. In fact, for the first decade of the Japanese cellular phone, there was little to remark upon. Indeed, it paralleled developments in an average European country, for example Spain, where a monopolistic public telephone service introduced an early but expensive system, but take-up rates were low. Nippon Telegraph and Telephone (NTT), then a public corporation, launched a cellular service around Tokyo, and a smaller one around Osaka, in 1979 – among the first commercial cellphones in the world. But ten years later, only 0.15 per cent of the population had bought the deal.

Of course, the peerless Japanese electronics manufacturers, such as NEC, Matsushita (under the Panasonic brand name) and Sony, supplied equipment to cellular services growing elsewhere in the world. But despite the excellence of the product, export suffered from measures taken to protect American or European markets. By the late 1980s many telecommunications markets were being opened up to competition. However, regulation could take protectionist forms, even if it had been designed for other reasons. In the case of GSM phones, the thorny and complex problem of patents meant that only a handful of companies (of which only Motorola were non-European) gained access to a highly profitable pan-European market. Japanese firms lay outside the walls of 'fortress Europe' for phones. They were also hampered by the bursting of the Japanese bubble economy in the early 1990s, which meant that the manufacturers were distracted, inward-looking and cautious, just as GSM was launched in Europe.

The Japanese mobile gets interesting – and very distinctive – from the late 1980s. NTT's monopoly was broken when two companies offered cellular services: the Nippon Idou Tsushin (IDO) corporation around Tokyo, and Daini Denden Incorporated (DDI) around Kansai. The resulting patchwork of incompatible standards bears comparison with Europe before GSM or the United States even now. While GSM represented for Europe a remarkable experiment in coordinated technological action, this style of policy, involving close cooperation between government agencies and private corporations, was for Japan a familiar and successful model of technological innovation. The result was a new digital mobile standard, agreed by 1989 and licensed to three consortia, led by NTT, Nissan and Japan Telecom respectively. With compatibility and competition came fast growth: nearly 9 million subscribers (7 per cent of the population) by the end of 1995.

It was NTT's mobile division, later a separate company eventually called NTT DoCoMo, that benefited from an astounding transformation of the mobile phone, a change led by the users themselves. The company had already built up a large customer base for its cellular phones: 1 million subscribers in 1993 (the year of the digital launch), 10 million by February 1997 and 20 million only eighteen months later. (And the growth would continue: 30 million in 2000 and 40 million by 2002.) But in 1999, DoCoMo created a new service.

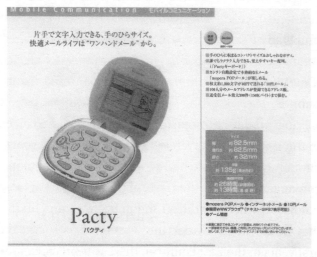

The DoCoMo 'Pacty', 2001. This device illustrates several distinctive features of Japanese mobile culture, in which image and design are highly valued: it is styled to appeal to the young female consumer, and greater attention is given to good graphical content. DoCoMo, a spin-off from the Japanese NTT, demonstrated the potential of consumer-led, data-rich mobile communications in the late 1990s.

Called 'i-mode', it was an instant hit. Distinctively, it was the hyper-fashion-conscious Japanese teenagers that led the way. Ten million had i-mode by 1999; a number that doubled in a year, and trebled in two. They wanted i-mode, but what was it?

I was amazed when I smashed my phone up how many components were inside. There's a lot more than just a microphone, loudspeaker and radio transmitter and receiver. Most importantly, there's a chip. The presence of a microprocessor means that a cellphone can potentially do anything that a computer can do: it can send, receive, store, show and change data in pretty much any form required. So there was never any reason to think of it as just a voice communication device. A mobile is – or could be – much, much more than a phone.

So a mobile could carry data as well as mere voices. In the 1990s, this was already old hat. But while speculators might pin their hopes – and investments – on a convergence of information and communication technologies (which produced the great telecoms bubble that would nearly wreck the world's economy in the first years of the 21st century), there was precious little evidence from anywhere in the world that consumers wanted more than a phone. Videophones, for example, in which you could see who you were talking to, have been technically feasible for decades. They have been launched several times, but always failed. Dashing the hopes of telephone companies, people preferred the low-tech, low-bandwidth voice-only telephone conversation, with its capacity to keep revealing expressions – and one's appearance – hidden. Text messaging had been written into the GSM standard, but almost as an afterthought. In the West, the attempt to combine limited access to online information via the mobile phone, based on Wireless Application Protocol (WAP), was greeted with much fanfare in early 2000, but was a disappointment. The limited range of WAP content available made the service seem tame compared to the wilds of the internet, and dispiritingly corporate. For those who wanted the corporate product of Disney, Nike or Champions League Football – and there were, of course, millions who did – there were better ways of consuming it than via the inch-wide WAP screen. For those for whom the restriction to corporate product was a turn-off? Well, they never turned WAP on.

i-mode was like WAP, but it worked. In both systems, the operator controlled access to digital content. But while the walled garden of WAP wilted, i-mode's flourished. To understand why, we need to look a bit closer at the horticulture: how i-mode content was grown and provided, and how it fitted with Japanese society. i-mode acted as a gateway. Any provider of content had to satisfy NTT DoCoMo's strict rules concerning what content was allowed and what was not. Content would be current, attractive and safe. This guarantee of quality meant that the owner of the i-mode phone would not be shocked or defrauded. In return for self-censorship, the content provider was granted access to NTT DoCoMo's already considerable market of sub-scribers. Furthermore, the complicated question of billing – the rock on which nearly all models of elec-tronic commerce are wrecked – was solved: amounts of downloaded data were totted up and charged as part of the phone bill. Only DoCoMo, gatekeeping transactions as both broker and guarantor, could do this. Furthermore, unlike WAP or most people's per-sonal computers (at the time), i-mode would be always online. Kei-ichi Enoki, managing director for i-mode, has summed up i-mode's philosophy of restricted con-tent and constant access:

> PCs are like department stores. They have a wide selection of content, including excellent graphic images. If you decide to make a visit, you can stay as long as you like and explore different sites at your leisure. [By contrast] mobile phones are more like convenience stores, where only a selec-tion of goods are on display in the limited space available. The contents have to be simple, but the convenience comes from the fact that they can be accessed at any time.

Convenience was of course a virtue in an intensely

hectic country such as Japan. Indeed, when i-mode was launched, the content sanctioned by NTT DoCoMo seemed squarely aimed at the busy salaryman. For ¥300 a month, a January 1999 press release promised subscribers the chance to reserve airline and concert tickets, check their bank balances, transfer money, send and receive email (a real draw, since government regulation of NTT had checked the growth of the internet in Japan), and to register for the 'Message Service', which automatically gathered information on 'weather, stocks and other topics, depending on choice'. All very safe, adult and probably heading for failure. In contrast, i-mode was discovered by teenagers and young adults. It particularly appealed to single women in their twenties, whose economic status was rising rapidly just as that of the working man with a job for life declined. A different, and very profitable, dynamic sprung up. i-mode not only gave its young subscribers instant access to new goods but also kept them in constant touch with each other, intensifying the pattern of rapid change still further. In turn, providers of commodities, well aware that the spending power of the Japanese teenager was second to none, had access through i-mode to a fast-moving, capricious but profitable market. The young yen saved i-mode, and demonstrated that the mobile phone could really be more than just a talking point.

Japanese mobile phones have remained highly

distinctive. Compared to American or European handsets, they will typically have far more specialised hardware features, come in a 'clam shell' design, and have more advanced functions more early. Mobile TV arrived first, for example, in Japan in 2005, and third-generation networks grew faster there than anywhere else. This mixture of advancement and distinctiveness is a source both of pride and concern for Japanese electronics companies. Firms such as Casio, NEC, Sharp and Panasonic sell well to their home market but very poorly outside. 'Japan's cellphones are like the endemic species that Darwin encountered on the Galapagos Islands – fantastically evolved and divergent from their mainland cousins,' explained Takeshi Natsuno to Hiroko Tabuchi in the *New York Times* in 2009. Natsuno is a particularly well-informed observer of the Japanese scene as he was one of the leading executives at NTT DoCoMo at the time of i-mode. This 'Galapagos syndrome' explains, wrote Tabuchi, why Japan's mobile phones have not gone global.

Now such a statement is true at the level of the design and manufacture of handsets. The global leaders now (in 2012) are firms such as Apple and Samsung. But there is one important sense in which the Japanese approach has anticipated, if not been copied. The 'walled garden' approach to content, with all its dangers, would be embraced by Apple throughout the app revolution of the 2000s.

Chapter 12
For richer, for poorer: India and China

China and India, such radically different countries, were often invoked together in the 2000s because they both possessed rapidly booming but uneven economies that, projecting forward, would soon rival those of the United States and Europe. They both had populations above, or soon to reach, 1 billion, and therefore also offered the largest potential markets on the planet for new products and services.

In 2004, whereas fewer than five in 100 people in India owned a mobile phone, this level of access already equalled the number of Indian fixed line customers, just over 40 million. What this tells us, of course, is that private individual possession of a telephone of any kind was relatively unusual on the subcontinent. The spread of the mobile phone has therefore been the expansion of private ownership at the expense of more public or shared means of communication. Indeed the growth in just half a decade was extraordinary: from 8 per cent of the population in 2005 to 70 per cent, reaching over half of all households, in 2010, according to the World Bank. And this expansion illustrates the continuing globalisation of cellphone businesses; Vodafone, for example, has seven times more Indian subscribers than in its home market, Britain.

Partly this growth was driven by the manufacturers, having saturated the markets in the West, slashing prices to chase the Indian consumer with offers of 'ultra-low-cost' handsets. (A similar pattern may be emerging in the 2010s, as companies such as Google work with Indian companies to develop cheap smartphones.) These ultra-low-cost handsets lacked many of the features of Western second-generation phones – screens would be black and white, not colour, and had few gizmos such as cameras or large memories. However, for the Indian consumer, as for the American, Japanese or European, the mobile phone was aspirational: a marker of status, albeit in complex and culturally embedded ways. The ultra-low-cost phone

An Indian farmer on his cellphone, just before the Baisakhi festival celebrations at the village of Pandori Waraich, near Amritsar. (Press Association)

therefore might have a market, but it would not be the desirable phone of choice.

Another issue is geographical coverage, not only of cellular networks but also of the internet, to which third- and fourth-generation phones need to be able to connect to be most useful. Even after great growth, and with eight companies competing for subscribers, the cellular networks only reached just over four out of five Indians in 2010. 'The bottleneck' in India, reports the World Bank, 'is coverage.' Likewise, for smartphones – phones which have rich visual interfaces and connect to the internet – the problem is access to the web. Indeed, in 2012, the BBC reported the case of an 'offline' mobile app shop that had opened in Mumbai to satisfy an unmet demand. This is related to another bottleneck, although in this case one of economic development: namely that the lucrative business of Indian programmers writing apps for the Indian market is being constrained by the lack of web access. As we shall see, however, not for long.

Modern India was formed by partition, and is bordered by Pakistan and Bangladesh. The geopolitics of cellular phone networks reflect the fraught relations between neighbours. In November 2009, following concerns expressed by the Indian intelligence agencies that untraceable phones were being used by militants, millions of mobile phones without valid IMEI (International Mobile Equipment Identity) numbers

– cheap, unbranded, ultra-low-cost phones – were blocked. On the border, in the Kashmir Valley specifi- cally, the Indian government banned prepaid phones, affecting over 3 million users. The ban was only lifted the following year, when a system of verifying the iden- tity of prepaid phone users had been hammered out. Once users became 'legible' – readable by authorities – the anxiety subsided somewhat. Tensions also arise over the fact that the radio waves sent and received by mobile phone masts are no respecters of international borders. Here for example is the *Times of India*, a rela- tively moderate paper, reporting on the issue in 2012:

> According to the reliable sources, Pakistan in its conspiracy against India has set up over three dozen new mobile towers near the international border adjoining Jaisalmer, Bikaner, Ganganagar and Barmer in Rajasthan. The signals of these mobile towers are getting into the Indian terri- tory up to 15–20 km, which is becoming a threat to the security for India. ... [Furthermore, many] smugglers and spies are getting Pakistan Sims ... in large numbers through the Thar Express.

Pakistani songs and speeches might be being smug- gled via this railway connection between Karachi and Rajasthan, while the newspaper also reported 'informa- tion that a few old smugglers in Longewala, Ghotaru,

Ramgarh area in Jaisalmer district, climb the old towers or go the [highest] places … [to use] Pak sims [to] send … confidential information to Pak intelligence agency ISI'. In other words, cellular networks near the border were seen as the tools of spies.

The spread of messages through the networks could also unnerve Indians, or exacerbate existing social conflicts. For example, India's north-eastern states are squeezed between Bangladesh and Myanmar to the south and China and Bhutan to the north. In this relatively isolated part of the country there is great ethnic and cultural diversity. When violence broke out in August 2012 in the north-east Indian state of Assam, between Bodo tribespeople and Muslim settlers, rumours of imminent attacks started spreading by text message and phone-based social media among north-east Indians living in the southern cities. Many thousands fled Bangalore, Chennai, Delhi and Mumbai despite repeated official reassurances, and indeed despite there being no evidence of an existing threat. 'Friends from NE please do not go by rumours being spread,' tweeted the Delhi police commissioner Neeraj Kumar: 'They are rumours and only rumours.'

In the case of China we have already seen how the march of the mobile phone masts mirrored the economic reforms as they gathered pace in the 1990s. In the mid-2000s, cellphones were relatively costly, and therefore the preserve of the more wealthy, middle-class

Chinese. People bought foreign brands, such as Nokia or Motorola, although there were Chinese mobile companies – Lenovo, Ningbo Bird, Konka and TCL – as well as so-called *shanzhai*, or black market, phones made by smaller, arguably even more entrepreneurial Chinese firms. By 2009, *shanzhai* phones made up one in five sales in China, and were being illegally exported to markets across the world, from Russia and India to Europe and the United States. 'We are a kind of illegal producer,' said Zhang Feiyang, owner of a firm called Yuanyang, to the *New York Times*. 'In Shenzhen there are many small mills, hidden. Basically, we make any type of cellphone.' One Yuanyang product was knock-off iPhones, made for a fraction of Apple's selling price.

By 2010 over seven in ten Chinese possessed a mobile phone. The billionth Chinese consumer switched on his phone in 2012, the same year as the Chinese smartphone market overtook the American.

Part three
Mobile cultures

Chapter 13
Txt msgs

Wot is kltr? Kltr is da clln of sgns spcfc to a sosIET. evry tek hs a kltr of its own. a kltr cn b hrd 2 undrstnd 2 outsdrs. ther is no bttr illstrn of ths thn txt mesgs

Txt msg ws an acidnt. no1 expcted it. Whn the 1st txt mesg ws sent, in 1993, by Nokia eng stdnt Riku Pihkonen, the telcom cpnies thought it ws nt important. SMS – Short Message Service – ws nt considrd a majr pt of GSM. Like mny teks, the pwr of txt – indeed the pwr of the fon – wz discvrd by users. In the case of txt mssng the usrs were the yng or poor in the W and E[1]

---

[1] In case you are not familiar with the text messaging 'language' commonly used, here is a translation:

What is culture? Culture is the collection of signs specific to a society. Every technology has a culture of its own. A culture can be hard to understand to outsiders. There is no better illustration of this than text messages.

Text messaging was an accident. No one expected it. When the first text message was sent, in 1993, by Nokia engineering student Riku Pihkonen, the telecommunications companies thought it was not important. SMS – Short Message Service – was not considered a major part of GSM. Like many technologies, the *power* of text – indeed the *power* of the phone – was discovered by users. In the case of text messaging, the users were the young or poor in the West and East.

Chapter 14
TxtPower

City life is mobile. City life is fast. And in no region did have cellular phones become so culturally important, so fast, as in the cities of the Pacific Rim. In entrepreneurial Hong Kong where, writes Garrard, 'it is almost as important to look busy and important as it is to make a deal', there has never been enough capacity to meet demand for mobile phones, despite licences being granted for any and every standard. Hong Kong needed six DCS 1800 (higher frequency digital phone) services, for example, when the United Kingdom had been satisfied with just two (Orange and One 2 One). When China took back Hong Kong from Britain in 1997, the mobile genie had escaped the bottle, and the strict controls over telecommunications found in mainland China would have been nearly impossible to introduce. Likewise, Australian cellular phone systems were centred on the flourishing cities, such as Sydney and Melbourne. Indeed, so strong is the expectation of being able to keep in constant touch via the mobile phone that inexperienced travellers into the outback have to be repeatedly warned that the cellular signal dies a few miles outside of major towns. But while the cellphone has bolstered the existing entrepreneurial culture of Hong Kong, and confirmed the sociability of

Australian city life, in the Philippines the mobile led to a political revolution.

Under Joseph Estrada, who had been elected president in 1998, many Filipinos felt that the country was slipping back to the corruption and cronyism of the Marcos days. But whereas the old dictator could introduce martial law and crack down on opposition, and the anti-Marcos coalition had to rely on ham radio and mimeographed pamphlets, the opponents of Estrada had a new tool in their possession. Mobile phones in the Philippines took a while to take off, mostly because spectrum space had to be cleared – a complex task when unofficial radio was a major means of contact in an island archipelago with patchy or absent telecommunications infrastructure. But in 1995 Smart Communications, a consortium that included First Pacific and NTT, entered the market, bringing much cheaper cellular service prices. By 1996, 7 million (10 per cent of the population) owned a mobile phone – almost twice the number with landlines. Perhaps more importantly, prepaid cellphone services were readily within reach of the poor.

Across the world a split had developed, by the late 1990s, between those who paid for mobile phones by monthly bills, and who therefore registered personal details and submitted to credit checks, and those – the young, the poor – who used prepaid services, usually by purchasing top-up cards, which had a lower starting cost (but often higher call charges), and who as a result remained anonymous. Text messaging was encouraged by the use of top-up cards, since to eke out the minutes it was better to use fractions of a second to send a text than waste whole minutes in conversation. (Indeed, texting was sometimes made free as an enticement to new customers.) In the United States text messaging was not popular, since phones were incompatible and the cost advantages mattered less to the affluent. (Plus, beepers and pagers had a prevalence unmatched elsewhere in the world.) As a result, mobile culture is far less rich in America than in texting hotspots such as Finland, Italy, the United Kingdom or, particularly, the 'text capital of the world', the Philippines. According to Rodolfo Salalima, the vice-president of lead Filipino carrier Globe Telecom, about 80 per cent of his company's customers used pre-paid cards, and the cheapest phone card cost about $5 and was good for two months. In an agricultural, Catholic country where the extended family was important, but also an industrialising country where the young were drawn or forced to the cities, $5 bought cohesion. The outcome was that by 2001 not only did the Filipino elite communicate by cellphone, but the rest of the population, vast and previously poorly connected, possessed anonymous text message-enabled phones too.

Text messaging played a key role in ousting Estrada. In late 2000, rumours spread as fast as fingers could text: true, exaggerated and imagined stories of Estrada's corruption. Over 100 million text messages flew around the Philippines each day. It started with jokes such as: 'The NPA [communist rebels] have kidnapped Erap [Estrada's nickname, which means 'buddy' backwards in Tagalog]. They are demanding a large ransom and, if it is not paid, they are threatening to release their hostage.' (To illustrate the indiscriminate power of text, another hoax announcing the death of the Pope was also passed on by millions.) As push came to shove, people were texted: 'edsa. edsa: everybody converge on edsa' – Edsa being the shrine that was the focus of the challenge to Estrada. While it was only after the cabinet had defected to the opposition, and the army and the police had transferred allegiance, that Gloria Macapagal-Arroyo was swept to the presidency in January 2001, it was also 'people power' brought together by text messaging that forced them to shift in the first place.

Macapagal-Arroyo acted straight away to ban 'malicious, profane and obscene' texts, which offered some protection against her predecessor's fate, but she has not been allowed to forget the power of text. In September 2001, Smart Communications and Globe Telecom announced that free texting would be reduced. Immediately, Txtpower, a group formed by cellular phone subscribers, organised the sending of at least 1 million text messages to President

Macapagal-Arroyo starting the following day, urging her to intervene to save free text messaging and to take action to improve the alleged 'lousy' services of the phone firms. Likewise, when in May 2002 President Arroyo announced plans for a tax on text messaging – the country's public debt was $50 billion, or 70 per cent of GDP – Txtpower and sympathetic politicians reacted angrily. For them, free texting was equated with freedom of Filipino expression.

The story of the Philippines shows, once again, that mobile phones are moulded by the countries in which they are used, and help to shape the nation in return, but it also acts as a reminder of another theme of this book: the shift away from centralised, hierarchical modes of organisation towards decentralised networks. This was not driven by technological change, although new technologies, of which the mobile phone and the internet were prime examples, symbolised and supported it. Instead they reflected the great shift in models of governance – that is to say, the ways in which decisions were taken and acted upon by organisations, be they government, industry or NGO.

Chapter 15
Two organisations in the Congo

In early 2001, Mount Nyiragongo, one of Africa's most active volcanoes, erupted. The lava flow passed through the centre of Goma, a Congolese city on the edge of Lake Kivu near the border with Rwanda. In the 1990s it had been the temporary home of the Rwandan Hutu refugees, many responsible for genocide, as well as the centre for the Rwandan-backed rebellion against Laurent Kabila's Congolese regime. Many who were part of the second wave of refugees, leaving Goma to escape from the advancing lava, recalled the miserable experience of the Hutus, and were determined to stay in the camps for as short a time as possible. Around Goma, therefore, two styles of organisation mattered.

The first was represented, in the absence of the distant Kabila government, by the United Nations High Commission for Refugees and the charities. These were organisations with clear centres, distributing aid along chains of command. The second was formed of the centreless networks of gossip, given technological form by the mobile phone. As CNN reported: 'An unlikely adversary has emerged in the battle to bring relief to the victims of the Congo volcano tragedy – the mobile phone.' Oxfam worker Rob Wilkinson said that while aid agencies were telling people not to return to the city of Goma and to stay in the refugee camps, mobile phone calls were persuading them to return. 'They are using mobile phones to talk to relatives and friends back in Goma, who are telling them that it is OK to go back,' he told the Press Association. 'It is changing the way the population is responding. It's very unusual.'

Interestingly, what can be a problem for the aid agencies – the self-organisation of mobile populations – can also be turned into a helpful tool. Following the devastating earthquake in Haiti in 2010, many of the population of Port-au-Prince fled the capital city. But relief agencies needed to know where essential supplies, such as fresh water and medicines, should be directed. Swedish and American researchers realised that mobile signal traffic could be quickly and cheaply traced to draw a picture of human movement. While it was too late to be of use in the aftermath of the earthquake, this tool was ready when an outbreak of cholera occurred ten months later.

Chapter 16
M–Africa

The changes, opportunities and challenges sparked by the introduction of mobile phones in sub-Saharan Africa provide some of the most extraordinary stories in the global history of the cellphone. Typically, mobile phones were introduced in a context where existing landlines were expensive, exclusive and often unreliable. In Nigeria in the late 1980s, for example, a single national telecoms company, NITEL, provided a mere 500,000 landlines for a population of over 100 million people. Early analogue mobile phones were equally restricted to an elite. When GSM digital mobile phones were introduced, in the early 1990s, possession rocketed, reaching 7 million subscribers by 2004. By 2011, six in ten households in a population of 160 million had access to a mobile phone, reflecting a similar proportion of subscriptions.

This pattern, to a greater or lesser extent, can be seen across Africa. Between 1998 and 2003, with the roll-out of African national GSM networks, often in competition with each other, mobile phone networks grew by 5,000 per cent. This growth rate was higher than the global average. In South Africa, by 2010, there were comfortably more mobile phone subscriptions than there were people, a penetration level greater

A buyer uses two mobile phones as he prepares to conclude a deal on a camel at the livestock market in the desert village of Sakabal, Niger, in 2012. (Press Association)

than that found in France. In some countries where there has been a particularly weak central authority, this growth has been particularly rapid. In Somalia, also by 2010, for example, about a third of the population had a mobile phone. Even among the countries that global authorities such as the World Bank and the International Monetary Fund list as the poorest (per capita) in the world, mobile phones have been purchased and used in surprising numbers: 14 per hundred of the population in the Democratic Republic of Congo, 41 per hundred in Liberia and 60 per hundred in Zimbabwe, to give 2010 figures. The demand has been intense enough to cause conflicts. For example, in November 2004, Conakry, the capital of Guinea, was shaken by riots as customers took out their frustration on the state telephone company, Sotelgi, when a promised delivery of new SIM cards failed to reach them, having been snatched up by middlemen.

Mobile phones are cheaper to use than landlines in Africa, which is part of their attraction. However, if we take Tanzania in 2005 as an example, a mobile phone might cost $50 to purchase and a call would cost 30 US cents a minute, while many Tanzanians lived on less than one dollar a day. Most of the money went to the mobile phone company (African Vodacom, based in South Africa, was the leading company in the case of Tanzania), but the fee might also include a chunk that went to the government in the form of a revenue surcharge. Mobile phones, as we all know, are tempting things to overuse. In Nigeria, the wry local name for a phone is *oku na iri ego*, which means 'the fire that consumes money'. The cost, especially of voice communication, has been therefore a substantial concern, especially for poorer consumers. This stark economic fact has shaped distinctive patterns of mobile phone acquisition, use and disposal across sub-Saharan Africa.

While brand new mobile phones are, of course, available for purchase in African cities and towns, there are also well-developed networks of redistribution which circulate old mobile phones from Europe, the Far East and the United States for resale. Jon Mooallem traced and described some of these networks in the *New York Times* in 2008. Most old American cellphones are just thrown away or left in a drawer. Some, beyond repair, might be sold to industrial recyclers such as the Belgian company Umicore, which once specialised in extracting the mineral wealth of the Congo but now smelts down 'e-waste' (electronic device rubbish) in an enormous facility in Antwerp, skimming out the valuable metals, including gold.

But many other phones, unwanted in the United States because they are months out of fashion, are fed into the chains of redistribution and resale. One route involves Pakistani wholesalers based in Kowloon, Hong Kong, orchestrating the movement of containers full of phones and shipping them from the United States to China. Nigerians then buy the phones in batches of tens of thousands and bring them to Lagos. From there the phones are passed out to sellers in Nigerian streets and markets.

Once fitted with new local SIM cards, the phones are ready to use. An African mobile customer needs three things: a phone (still expensive, even when second-hand, but often, as we will see, an investment), a SIM card, and a way of paying for calls. Because of the high cost of calls, many Africans – just like the Filipinos earlier – choose pay-as-you-go tariffs, in the form of top-up credits bought from street kiosks. Pay-as-you-go is cheaper but also, crucially, more manageable.

A kiosk in Nairobi, advertising Celtel phone services alongside its other product lines. (Press Association)

Another strategy to keep costs down is to share a phone. Along with the kiosks selling pre-pay vouchers, a familiar sight in contemporary African towns and cities is the brightly-coloured call box, where a local entrepreneur bulk purchases and resells time on cellular phone networks to customers, who are charged by the minute. One of the most important models was an extension of a micro-credit system that started in Bangladesh. The Grameen Bank (which won its organisers the Nobel Peace Prize in 2006) provided credit and facilities for first Bangladeshis, and then Ugandan and Rwandan women to set up 'Village Phones'. In Tanzania, a similar shared call box run by a local entrepreneur is known as *Simu ya watu*, which translates as the 'People's Phone'. The outlay is daunting, but the rewards and effects are there. 'Mostly we have fishermen here, we have farmers and we have the business community ... people from other countries and other districts,' the local call box entrepreneur of Kigoma, Mwilima Ahmed Kalunga, told Jon Cronin of the BBC, adding 'It cost $13,000 [from Vodacom]. But most people like to use us because they can see the minutes tick. They cannot be cheated.' Sharing a phone can be shaped by cultural as well as economic factors. The anthropologist Daniel Jordan Smith, for example, found that mobile phone credit was perceived as more akin to food and drink, which you are expected to share without incurring a debt, than to money. In this case, a new form of credit was being grafted onto existing expectations of reciprocity, gift-giving and sharing.

Another common practice prompted by the economics of African mobiles is the widespread phenomenon of 'flashing'. Since a call is usually not charged if the recipient does not answer, letting the phone ring once and hanging up signifies that a speaker would like a return call. The return call will, of course, be at the second person's expense. Since most people are down on credit sometimes, the system allows the expense of calling to be shared, so long as narrow self-interest does not dominate. In Khartoum, in Sudan, tea-sellers can be summoned through this missed call system. But its application is usually broad and reciprocal. Smith, who recorded this practice in Nigeria, notes that the recipient of the flash does not always ring back. Indeed, much like voicemail, flashing provides the facility to screen as well as manage costs. In a culture, like the West but perhaps more so, where phones are status symbols and flamboyant visible phone use can be a form of conspicuous consumption, and when (and for how long) phones are used can indicate, as well as maintain, social relationships. Nevertheless, 'flashing' works well between peers, although Smith also occasionally witnessed a 'comic exchange [of repeatedly flashing and flashing back] in which no one wants to bear the costs of a call'.

The mobile phone is used in Africa in diverse ways, some of which are similar to patterns elsewhere, while some are distinctive and innovative. Inge Brinkman and her anthropological colleagues Mirjam de Bruijn and Hisham Bilal have been observing the use of mobile phones for a decade. In Khartoum, mobile phones are used to sustain both economic and social ties. Young men, having moved to the capital, keep in touch with home. Family ties are sustained by the increased contact between kin. Young women, scandalously to some in Sudan, chat about relationships with men, and vice versa. Not only businessmen but also businesswomen, significantly in an Islamic society that enforces separate spheres of activity for men and women, use their phones to keep in touch with customers. The anthropologists give the example of Fatima, a henna painter:

> All her customers reach her by phone and she used the first income she had (in 2002) to invest in a mobile phone. When asked the reason, she answered: 'I heard the mobile phone would bring work and that was exactly what has happened.'

In countries such as Rwanda, Uganda and Tanzania farmers use their phones to access market information, such as tomato prices. Whereas previously they had been isolated and reliant on the say-so of middlemen, now they can check what a good selling price might be, and even arrange to sell direct. In South Africa, by 2005, more than eight in ten small businesses run by black people relied solely on mobile phones rather than landlines. The opportunities for entrepreneurship have been greatly extended by the spread of the mobile phone networks, not least because Africans have been able to carve out new jobs servicing them. The kiosks selling phone vouchers and the call boxes offering shared use of phones are both widespread examples. The general economic boost given by this activity has been measured. The London Business School, for example, reckoned that an increase of ten phones for every hundred Africans would boost GDP by over half a per cent. There are also suggestions, albeit slightly reminiscent of 'Darkest Africa' stereotypes, that the increased feelings of safety and security enabled by allowing businessmen and women to be in constant touch boosts the likelihood of trade between African hinterlands and the developed world.

Perhaps the most spectacular, and innovative, development of mobile culture in Africa is M-PESA – 'M' for 'Mobile' and 'pesa' meaning 'money' in Swahili – a form of cash carried on a cellphone that has been immensely successful in Kenya. Mobile credit has been tried in the West but it has never flourished. However, in Africa the different economic and social circumstances created a very different context, one conducive to growth. Banks might be only in large towns and cities (and they charged considerable fees), while credit was almost impossible to secure for those on the poorest incomes, and paper money, carried on the person, was vulnerable to theft. In 2004, M-PESA began as a pilot project, jointly funded by the UK Department for International Development and Safaricom, a Kenyan mobile company affiliated to Vodafone.

Interestingly, Nick Hughes and Susie Lonie, the organisers of the pilot project, conceptualised M-PESA as a microfinance loan repayment system. It was the Kenyan customers who discovered for themselves the extraordinary array of applications. The mobile wallet is therefore an African innovation. M-PESA was launched in March 2007. The set-up was simple: a

Kenyan registered with a local agent on production of an ID card and then deposited some credit on their phone account. This credit could then be used to pay for anything from bills to beer. Even more importantly, money could be transferred to another person's phone by text message. It was handy, safe, and scalable: small payments were as easy as large ones. As Safaricom's upbeat Kenyan website advertised it:

> From the 'mama mboga' in a village somewhere to a business magnate closing deals in a lavish hotel in the city, M-PESA is the first thing that comes to mind when a need for a financial transaction arises. Whether it is paying bills, buying airtime or avoiding the long inconveniencing queues in banks and utility paypoints, M-PESA has the solution for you. So the next time a financial need arises, sit back, relax and bask in the peace that comes with the reassurance that you have got M-PESA.

By 2009, nearly 7 million Kenyans had registered, and an equivalent of $2 million was being transferred every day. Most of these transfers were for small amounts, the kind of credit that would have been impossibly expensive for most Kenyans in the old world of bank accounts.

Anthropological work carried out for CGAP, an

A queue forms at an M-PESA office, Nairobi, Kenya, 2011.
(Press Association)

offshoot of the World Bank, by Olga Morawczynski has revealed how M-PESA is used in Kibera, a slum of 1 million inhabitants on the outskirts of Nairobi and in Bukura, a farming village in western Kenya. Typical users, she found, were men working in the city, who sent money; and women in the countryside, who received money. The impact on rural recipients was dramatic; their income increased by a third. Partly this was because they no longer had to travel – a waste of time and money – in order to collect old-style transfers from towns. Urban users chose M-PESA because it was cheaper, faster and safer than alternatives. M-PESA was popular both with those with bank accounts and those without.

However, there were also problems and unanticipated outcomes. Congestion on the Safaricom network meant that sometimes M-PESA would stop working, leaving customers angry. Rural wives worried that their husbands, who because of the ease of M-PESA were making fewer visits home, might 'become lonely and find a "city wife"'. 'This finding counters a popular assumption that is often made about mobile phones – that these technologies amplify existing relationships,' writes Morawczynski and her co-author Mark Pickens, adding: 'When used as tools for financial services, these technologies can have the opposite effect.'

The anthropologists also noted that the flow of credit could go into reverse during times of crisis. In December 2007 a presidential election was followed by an outbreak of violence, as opposition parties contested the reinstatement of President Mwai Kibaki. Ethnic tensions rose to the surface. Running battles took place in the Nairobi slums, railway lines were torn up, the police shot dead protestors, and a church was set alight, killing 200. Tens of thousands of Kenyans were displaced. The unrest lasted well into the spring of 2008. The M-PESA credit sent home to the countryside was returned hurriedly to the men in the towns, who reconverted the credit to money to escape the violence, or used it on their phones to pay for communication.

In general, then, African mobile has displayed two outstanding features: growth and innovation. By the end of 2012 the estimated number of subscribers on the continent was 735 million, over six in ten of the population. This leap was from pretty much a standing start in the 1990s. The innovation has been locally led – M-PESA as an all-purpose mobile wallet is perhaps the most striking example – but it is matched in inventiveness by a host of smaller-scale mobile technology ideas, from devices to detect shoals of fish to ways of tracking stolen vehicles.

Chapter 17
The Nokia way – to the Finland base station!

The case of the Goma volcano reminds us that a key aspect of mobile culture is, perhaps obviously, mobility. (And a mobility that cuts across national boundaries.) But there is also a distinctive *material* aspect to mobile culture, which is best illustrated by the products of the phenomenal Finnish company, Nokia. While material culture might seem at first to be mere flotsam and jetsam, and not part of the great tides of history, I think the opposite is often the case. Indeed, something as trivial as a coloured plastic phone cover – called a facia or fascia – can arguably be as much a vehicle of grand historical change as fascism.

Nokia became, without doubt, the most influential manufacturer of mobile phones in the world. But why did such a firm emerge from Finland? Industrialisation came late to this country on the northern fringe of Europe, but when it did so it made use of one natural resource Finland, like Sweden, had in abundance: forest. In 1863 Knut Fredrik Idestam, after a daring act of industrial espionage, imported a new wood-pulp process from Germany, and set up a mill on the Nokia river, which flowed a few miles outside the small city of Tampere. Right from the start the enterprise had close links with Finnish politicians: Idestam's partner was Leo Mechelin, a parliamentarian and financier who helped extricate Finland from its status as a Russian duchy to being an independent state. For much of the 20th century Nokia was an industrial coalition between pulp, rope, cable and rubber works. Indeed as late as the 1980s, as Nokia's historian Dan Steinbock records, Nokia at one and the same time brought electricity to 350 Egyptian villages, made most of the toilet paper in Ireland, and provided all the studded bicycle tyres in the world.

But the peculiar position of Finland in world politics meant that Nokia was quite unlike any other European conglomerate. Firstly, ever since the Russian revolution, Finland had had to play a delicate balancing act between capitalist West and communist East. This strategy, a combination of cautious neutrality and *realpolitik*, has a name: the 'Paasikivi-Kekkonen' line, named after the two politicians who adopted it. For example, Nokia's major market was the Soviet Union (not least supplying much of the power cables for Lenin's, and later Stalin's, programme of electrification), but later, with some prescience, Nokia's boss Kari H. Kairamo decided that such dependency should be balanced by building up western-European links. The policy had been echoed at a national level; Finland, after delicate negotiations between East and West, joined the European Free Trade Area in 1961 and signed trade agreements with the European Economic Community. Second, Finland was extremely dependent on imported oil; this had to be paid for by increased exports, which again gave cause for good trading relations with both East and West. So, under Kairamo from 1977, Nokia sought the means to create innovative exportable electronic products. Finally, when the telephone originally came to Finland, it was not – unlike in any other country in Europe – placed under the control of a single monopolistic operator, but was controlled by a host of independent local operators instead. There were over 800 in 1938, and still around 50 in the 1990s. Again, the cause can be found in Finnish foreign relations. The distinctive Finnish telecoms pattern of links between private companies and many local cooperatives was, notes Steinbock, due not so much to 'boosting the efforts of the private sector as trying to keep Russian authorities away from the emerging industry' (a strategically important industry at that). The important consequence for Nokia was that it had on its doorstep a diverse market for competitive products, and that it never had to compete with a big national telecoms monopoly.

Nokia had already manufactured a few mobile phones, at its Oulu plant in the far frozen north of the country, when Kairamo signed deals with the Finnish TV manufacturer, Salora Oy. The joint venture, Mobira Oy, begun in 1979, was soon owned outright by Nokia when it swallowed up Salora in 1984. Also in this period, and for reasons that are unclear, Kairamo tore down Nokia's hierarchical organisation, typical of many a European conglomerate, and replaced it with a decentralised 'flat-pyramid' management. This radical change, which would only later become a new orthodoxy of managerial science, seems to have been based on Kairamo's shrewd analysis – or guess – that the world system of two superpowers was nearing its end and only a nimble company would be able to exploit the new global opportunities. What is certain is that if the tremors of the coming earthquake *could* be felt, then Finland, balanced precariously between East and West, was near the epicentre.

Kairamo, who had manic depression, hanged himself in December 1988, having become convinced that a forced break-up of Nokia was imminent. In fact the restructuring that later took place, initiated by Simo Vuorilehto and completed by Jorma Jaako Ollila, built on Kairamo's legacy. Ollila was the person most responsible for focusing Nokia almost entirely on mobile phones. In effect Mobira Oy, along with a few other electronics and cable divisions, became the whole company. Many factors had combined around 1990 to permit this. Finland had a purely conservative government, intent on telecoms deregulation, for the first time in decades. The Finnish economy was in tailspin following the collapse of Soviet trade, and a new direction was needed. This direction clearly pointed towards further European integration (Finland joined the European Community in 1995), of which GSM was to be the showcase for pan-European potential. Indeed the pro-European emphasis, echoed at a national level, had already prompted Nokia's involvement with the Nordic NMT standard and in early GSM discussions. So Ollila bet the company on mobile phones.

Jorma Jaako Ollila, chairman of the board and CEO, Nokia. (Nokia Photo Archive)

But if placing Nokia in a political context helps us understand why it was in a position to become a major mobile phone manufacturer, we need to go a bit further to account for its extraordinary success based on distinctive products – the material culture. Nokia's mobile phones of the 1980s, such as the Cityman (1986), which was very popular in the UK, already boasted superior design. Styling and brand were more important to Nokia than they were to competitors – such as Motorola or Siemens, say – or at least were pursued with greater success. (Again there was a cultural advantage: good industrial design was the material analogue of Nordic social welfarism – sharing the benefits of industrial society through rational planning.) But let's take just one object – to my mind, an iconic one – and look to see what it shows about the Nokia way.

The Nokia 3210 was launched in summer 1999. It is a design classic. 'Elegantly styled, with no protruding aerial and lovely slim proportions,' drooled *What Cellphone*. I, too, recall the thrill: the silver and grey phone seemed moulded to fit perfectly in the hand. It was obviously, immediately an icon, like Coca-Cola's bottles or the Citroen DS. (And by 2020 the sight of a Nokia 3210 will trigger millennial nostalgia.) In one sense the 3210 did for mobile phones what the Model T Ford did for the automobile: it was a cheap, but beautifully engineered, vehicle for mass communication. But if the Model T Ford symbolised the dominant style of production of much of the 20th century, the Nokia 3210 represented its opposite. Fordism stood for centralised control, hierarchical management and, famously of the Model T, 'any colour, as long as it's black'. Nokia boasted flexibility of production, flat hierarchies and products that reflected this organisational style. With the Nokia 3210, you could change its colour simply by choosing a new 'Xpress-On' fascia. *What Cellphone* relayed the reaction of Janice Caprice, a London beauty therapist:

> It's got to be eye-catching, anything from the British flag to a flower. Most of my friends buy a phone because they can get a cover for it. I bought an Ericsson PH337 for that very reason but that's old now so I'm saving up for a Nokia next.

The Nokia 3210 was to cellular communications what the Ford Model T was to the automobile. (Nokia Photo Archive)

Nokia had experimented with Xpress-On with the slightly earlier introduction of its more conventional 5110 phone to market in 1998, and other manufacturers had aped the innovation. But with the 3210, interchangeable fascias became integral to the product's design and marketing. Fascias are superficial and shallow. But they are also colourful and flexible, and mean that the same phone can display different allegiances, as fashions shift. I think we should take such superficiality seriously. The 3210, like the 5110, carried simple games derived from earlier classics (Snake, Memory and Rotation), more evidence of the incipient shift from mobiles as mere communicating devices to something more. The 3210 was also the first Nokia phone to carry T9, a predictive text system developed by a small company called Tegic, which shoehorned the equivalent of a full alphabetic keyboard onto just a number pad: the phone contained a dictionary and software for searching it, possible only because the mobile contained a microprocessor.

The contrast between the squat black Type 300 Post Office phone from my childhood and the chameleon-like 3210 should be clear. To hold the 3210 in the palm of your hand is to have evidence, in material form, of a great transformation.

However, the mobile world does not stand still. By the mid-2000s, Nokia's lead in stylish design was being eroded and lost. The launch of Apple's iPhone in 2007, and the subsequent redefinition of the 'smartphone', discussed in detail later, caught the Finnish firm on the hop. The challenge had been expected to come from Microsoft. In 1998 Nokia had teamed up with other phone companies, as well handheld computer pioneer Psion, to found Symbian, a company intended to develop operating systems for cellphones that would see off the Seattle giant. By 2004, Nokia had bought out most of the other stakeholders and had taken control of Symbian (full buyout occurred in 2008). Nevertheless with Nokia's sales on the slide, Jorma Ollila stepped aside as chairman and chief executive in 2006.

Chapter 18
Mobile phones as a threat to health

Stories of bodily harm caused by mobile phones were commonplace across the industrialised world by the late 1990s. The stories were directed against two culprits – radiation from base stations and radiation from the mobile phones themselves – but the tones were very similar. A story that could be found in the *Daily Mail* in December 1999 was typical. Under the headline 'Now mobiles give you kidney damage', the reader was told that 'scientists say exposure to the phones' low-level radiation causes red blood cells to leak haemoglobin. The build-up of haemoglobin, which carries oxygen around the body, can lead to heart disease and kidney stones.' The reader would already have known of earlier stories suggesting links between mobile phone use and brain cancer, premature ageing, diseases such as Alzheimer's and Parkinson's, multiple sclerosis and chronic headaches.

Chilling stuff. And although there was the usual disparity between headline ('mobiles give you kidney damage') and research ('more work is needed to investigate some results which seem to indicate that electromagnetic waves in the radio spectrum may interfere with processes within the kidney'), the economic

A base station aerial being erected in 1985. Two components
of cellular phone systems provoked anxieties over adverse
health effects: the handset and the base station. Base
stations near – or on top of – schools caused the greatest
concern. (BT Archives)

importance of the mobile industry forced governmental organisations to act. In the United States, regulation of cellphones is shared by the Federal Communications Commission (FCC), which sets guidelines concerning levels of radio frequency (RF) radiation, and the Federal Food and Drug Administration (FDA), which has a brief to follow health matters. The FDA set out to reassure cellphone users that the technology was safe.

In Britain, the Department of Health played a similar role to the FDA. However, with fiascos such as BSE in the recent past, the government chose to ask an Independent Expert Group on Mobile Phones to investigate. The group reviewed media coverage, and from September 1999 heard evidence from scientists, members of the public, representatives of the telecoms industry and special interest groups, such as Powerwatch, Friends of the Earth Scotland, and Northern Ireland Families Against Telecommunications Transmitter Towers. The findings, called the Stewart report after the group's chairperson, the biologist Sir William Stewart, were published in 2000. The balance of evidence suggested that exposure to radio frequency radiation at levels below existing guidelines 'did not cause adverse health effects'. However, the Stewart report went on to say that 'there may be biological effects' at such levels, and therefore it was 'not possible at present to say that [such] exposure ... is totally without potential adverse effects, and that gaps in knowledge are sufficient to justify a precautionary approach'. In particular, it said that children should not be encouraged to use mobile phones because their bodies were still developing.

The Stewart report's conclusions were more cautious than those of other governments' investigations. The Health Council of the Netherlands, for example, concluded that there was 'no reason to recommend that mobile telephone use by children should be limited as far as possible'. But such reports were also shy of making strong general claims over the safety of mobile phones. The World Health Organization (WHO) on the one hand records that 'none of the recent reviews have concluded that exposure to the RF fields from mobile phones or their base stations causes any adverse health consequence', but on the other felt the need to rush out statements correcting press articles which reported that the World Health Organization had insisted 'mobile phone emissions are safe'. The billion-dollar insurance claims over damage caused by asbestos and tobacco have made all organisations wary of putting their name to pronouncements of complete safety.

In the 2000s there have been further sporadic claims about the health risks of cellphones. In 2002, Finnish scientists claimed that the electromagnetic radiation affected brain tissue, while Swedish counterparts pronounced a link between users of early phones and incidence of brain tumours. A German-led European laboratory study using mouse models announced in 2004, to some alarm, that mobile radiation could cause genetic damage. In 2006, a British researcher at the University of Staffordshire linked mental wellbeing issues, such as stress, to mobile use. In response to this background there have been many attempts to close the debate over the health effects of mobile phones. But, as in other controversies such as that over BSE, experts tend to disagree rather than agree. The issue is also necessarily open-ended, since it is impossible to say what length of time will be enough for scientists to be satisfied that long-term harmful effects do not exist. Nor will a technical fix soothe fears. A small industry has grown up offering technical solutions, from headsets (so the phone irradiates your guts rather than your head), to fraudulent quack remedies involving 'absorbent' phone covers. These products exist *because of* anxieties, not to allay them.

Rather than expect the debate over health and mobile phones to be resolved, we should consider two quite different ways of thinking about it. First, with subscription to mobile phones hitting over three-quarters of the population in many countries, the big picture is one of users not resisting a technology, but enthusiastically embracing it, despite knowing there may be risks. What needs to be explained is not so much why there is public concern over harmful effects of mobile phones, as why the concern has so little effect on behaviour. Second, the debate will never be closed by expert pronouncement, since public concerns are framed by a powerful and growing culture of distrust towards scientific expertise. (I suggest that this trend is part and parcel of wider social and political transformations, discussed later.) Concerns supposedly directed at a particular technology are in fact generated by deeper social tensions and conflicts. Indeed, expert opinion has only united to declare harmful effects in one area: talking on a mobile phone while driving. This opened a new attack on an old alliance between two technologies of mobility.

Chapter 19
Cars, phones and crime

New technologies of mobility create new crimes, new criminal *modus operandi* and new ways to catch criminals. Take, for example, the automobile. From the first decade of the 20th century, cars provoked a crime wave. In Britain, a speed limit of 20 miles per hour was in force between 1903 and 1930, when it was briefly revoked; and a 30 miles per hour limit in built-up areas was hurriedly reintroduced by the 1934 Road Traffic Act, after a spate of accidents, and has been in place ever since. Mobility increased through the century: by 1950 an average day's travel was five miles, but by 2000 it was 28 miles. Much of this travel was by car. A dramatic effect of the new mobility, with its legal limits, was to create new crimes of speeding. Criminal statistics became dominated by car-related crime. Furthermore, as well as driving infringements becoming criminal offences in their own right, speedy automobiles led to more accidents, including 'hit-and-run' incidents. Patterns of criminal behaviour also changed. More cars meant more stolen cars. Burglars, previously confined to large towns and cities, suddenly found rich pickings in the surrounding countryside, which was now only minutes away. Statistics for break-ins correlate strongly with increased car ownership.

The history of two technologies of mobility and individual freedom, the car and the mobile phone, has long been intimate. (BT Archives)

But, in turn, techniques of policing and crime detection also changed in response to the mobility of criminals. On a mundane level, many hours of police time were now spent in work such as enforcing traffic codes or tracing missing vehicles. This new emphasis would have been unwelcome were it not for the new information tools that registration provided. Within a few decades of the birth of mass motoring, every driver and every vehicle had been tagged with a unique identifying number, and immense filing systems recorded changes and movements. The registers relating to motoring soon became central tools in police work. Of course, different countries responded to the mobility provided by the automobile in different ways. Norway, for example, introduced strict speed limits relatively early – a pattern shared by other countries which strongly valued social cohesion and were not swayed by lobbying from car manufacturers. The United States or China would be different cases again. But the car became so central to many societies in the 20th century that the phenomenon of increasing interdependence of policing and criminality was a common one. It should be seen as one more example of how the increased spread of technologies of mobility created possibilities of freedom from centralised authority – whether it was from the family, society or other enforcers of 'correct' behaviour – which in turn were countered, even domesticated, by strengthening bureaucratic or policing powers.

A technology that demonstrated the links between mobile radio and the automobile, between cultures of mobility and freedom, was Citizens' Band (CB) Radio. CB was not a cellular radio service: all transmissions were made on a segment of the radio spectrum, around 27 MHz, that had been given over to amateur use. Everyone could hear everyone else. From the late 1950s, a movement of enthusiasts was fostered by the marketing of CB radio kits. But CB radio really took off in the late 1960s and early 1970s, when its adoption was spearheaded by US truckers, providing communication and social ties across the freeways. The CB 'fad' marked the moment when the values – and technology – of the truckers struck a deeper chord. For a few years in the mid-1970s, the specialised CB language of American truck-drivers – '10-4, good buddy', 'Breaker 1-9, this is the rubber duck' and so on – could be heard echoed as far from Route 66 as Britain and the Netherlands. The 1977 film *Smokey and the Bandit* was a smash, as was *Convoy* (starring 'Rubber Duck') the following year. What resonated was the mythology of a network of individuals living outside traditional society, indeed outside the law, best encapsulated in the supposed truckers' trick of using CB radio to warn each other of police speed traps. The network opposed centralised power. While the popularity of CB prefigured the later demand for cellular telephony, the continuity of social values associated with both also tells us something important.

The cellular phone is also a technology of mobile communication, and a similar story can be told. As with the spread of the automobile, this new technology of mobility created new crimes. In early-21st-century

Britain, for example, the theft of mobile phones contributed to the largest recorded increase in reported crimes affecting young people. By 2007, 800,000 phones were being stolen in Britain each year. Rates were similar elsewhere: hundreds of thousands in Australia and millions in the United States. As we have already seen, the mobile phone is by far the most expensive object ever to become routinely carried on the person. (And by 2001 70 per cent of adults and 81 per cent of fifteen- to 24-year-olds in the UK possessed one.) Also, despite the phone's slide down the social scale, its role as an indicator of status and personal identity – important, as ever, in the playground – has not been diminished. But by 2001 over a quarter of all robberies involved a mobile phone, and nearly half the victims were under eighteen. The perpetrators too were likely to be young.

A portion of this wave of thefts by and from children stemmed from bullying rather than anything more organised. But when Home Office researchers Victoria Harrington and Pat Mayhew interviewed the inmates of Feltham Young Offenders' Institution, a more complex picture emerged. Some phones were stolen as part of the general 'pickings' of others' possessions, but expensive models might be particularly targeted. Phones would be used until the service was blocked, or the pay-as-you-go card ran out. Or they would be immediately passed onto a ready local – and probably international – market. Thefts were prompted by opportunity and, intriguingly, the social tensions brought to the surface by conspicuous display of technology:

> The plethora of phones around – left for instance casually on counter tops in bars, or on the shelf near the front door at home – also makes them relatively accessible to thieves. In the case of street crime, too, potential thieves can easily spot someone with a phone. It is difficult to say how much truth there is in the contention of offenders in Feltham … that owners' ostentatious use of phones causes a degree of irritation that provokes theft – but it might give users pause for thought.

The Feltham boys expressed the same 'irritation' as the annoyed passenger on the train surrounded by mobile users, although they took rather more drastic action in response. Many police officers (although not the Home Office researchers) believed that mobile theft was often implicated in the maintenance of social status by the technique of 'taxing': here 'the phone theft per se is less important than groups of offenders exerting control, establishing territorial rights, and showing "who's who" by penalising street users (in particular young ones) through phone theft'. Possession of technology made a social statement, and, conversely, to *not* have a phone meant social deprivation: for the boys, phones were 'an indispensable crutch. Their loss of phones in custody was said to be one of the worst elements of the deprivation of their liberty.' Loss of a technology of mobility was equated directly with loss of freedom.

By the 2000s it was clear that stealing mobile phones had become international and more organised, partly because increased security meant that phones were hotter property. The National Mobile Phone Crime Unit in the UK, for example, in 2007, noted that phones were being stolen to order and then shipped abroad in increasing numbers.

Reliable research on criminals' uses of mobile phones is much harder to come by, although evidence suggests that the opportunities provided by mobile phones to make planning and carrying out crimes more 'efficient' have not been ignored. Smuggling drugs across the Mexican–United States border (as well as the more benign movement of immigrants) is coordinated by text messages containing information about routes and the presence of border patrols. The messages are sent by spotters who, according to Marc Lacey in the *New York Times*, 'monitor the southern Arizona desert from lookout points and help steer the migrants, as well as drug shipments, away from authorities'. The patchy reception in the thinly populated border regions is a problem both for the patrols and for those they are trying to catch.

A major influence on criminal uses has been the availability of pay-as-you-go mobile phones, which can be easily bought under an assumed name. Any calls made might be traceable, but ownership of the phone cannot be ascertained. The 7 July 2005 tube and bus bombers in London exploited this anonymity. The same advantage, of course, applies to stolen phones. Before the mobile phone, the crime of cellular fraud could not have existed – another reminder that new technologies of mobility *create* new crimes. The subscription fraud form of mobile crime, in which phones are bought under false names with no intention of paying for calls, is the most frequent. But the cloning of phones, in which the identity of another user is stolen, amounted to 85,000 cases in the United Kingdom by 1996, and is widespread across the cellular world: 80 per cent of drug dealers apprehended in the United States in the late 1990s, for example, were found to be using cloned mobile phones; while in the occupied territories of Israel, Palestinians made 57,000 international calls, subverting official security restrictions – and charged subscribers in Arizona.

In 2001 a complex European taxation swindle that proved lucrative came to light. It was based, rather bizarrely, in Stoke-on-Trent, a conurbation in the northern Midlands of England which had not seen the good days since Josiah Wedgwood produced pots there. Fraudsters exploited the fact that no tax was charged on the traffic of goods within the European single market. They imported mobile phones, chosen because they were easily portable and in high demand, and sold them on to other, legitimate traders – adding 17.5 per cent tax, which was pocketed. At least £2.6 billion was creamed off in this way in 2000–01, with possibly 10 billion pounds lost to the taxman overall.

Just as the mobility of the car was domesticated by keeping databases of information, the criminal potentialities of the cellphone have been contained by technologies of registration. Indeed, whereas there was no essential reason why cars and drivers had to be registered – if you don't care about the law, you can get in a car and drive off; if it has petrol it will work – the cellular phone is unimaginable without databases of information. Every time a cellular phone passes from cell to cell, a reference has to be made to some central computerised list of network users. Every time an owner is billed, the information on phone use has to be totted up and processed. Unlike the automobile, databases are integral to mobile phone technology. (Although the user might never suspect this, sometimes with unintended consequences: when itemised billing was introduced in France male customers complained in their thousands because their extramarital affairs were uncovered overnight. The networks were forced to bow to public pressure and replace the last four digits of phone numbers with asterisks.)

Most of this information is stored centrally, as part of the operation of the mobile switching centre, but increasingly personal data is held in the handset. In Europe, this shift coincided with the introduction of digital cellphones. First developed for the German analogue Netz-C system, which in many other ways was a commercial disappointment, the Subscriber Identity Module, or SIM card, was adopted as part of the pan-European GSM concept. While the SIM card contained information identifying the subscriber, it also could hold much more: preferred phone settings, recently dialled numbers or the equivalent of an address book. (One incidental consequence has been the reduced need to remember phone numbers. I can no longer recall friends' numbers, since my phone identifies incoming and outgoing calls by name.) The introduction of the SIM card was prompted by anxieties over crime. The SIM card also reflected wishes – hopes rather than expectations – that the card, which could identify the purchaser, would spark a boom in mobile commerce. GSM, remember, was imagined as a key infrastructural component of a future European market.

The GSM standard also provided the option of identifying equipment through an Equipment Identification Number (EIN) – so the SIM would say who was on the phone and the EIN would say exactly which phone was being used. In the UK, the mobile phone companies chose *not* to activate the EIN capability. In April 2002 the British prime minister, Tony Blair, made an announcement that many commentators considered a hostage to fortune: that the recent rise in crime would be under control by the following September. What seemed like a bold political gamble was in fact partly a measured guess that if the mobile companies were persuaded to activate the EIN, then stolen phones could be cut off as soon as they were reported. Crime and mobile phones had become matters of political spin.

Of much greater significance was the adoption of the International Mobile Equipment Identity (IMEI) code. While the SIM card would be tied to the person, the fifteen-digit IMEI number was tied to the device. Try it yourself. Tap in *#06# and, on almost all cellphones, your IMEI will be displayed. (Why not write it down now in a safe place!) If a phone is stolen then it can be blocked by using the IMEI to disable the SIM card. Unblocking – the changing of a IMEI number, quite different from the harmless 'unlocking' which frees a phone from being tied to one company's service – by anyone other than manufacturers is illegal in most countries.

Meanwhile the information held by mobile phone operators, and even the personal data stored on the SIM card, have become detective tools and forensic evidence. In 1993, the hiding place of the cocaine baron Pablo Escobar was found by tracking his mobile radio calls. In October 2002, similar surveillance probably contributed to the arrest of al-Qaeda suspect Abu

Qatada. In 2010 the chance eavesdropping of the cell-phone of Osama bin Laden's courier, Ibrahim Saeed Ahmed, led American investigators to the Peshawar district of Pakistan. Further intensive, and deeply secretive, probing located the al-Qaeda leader's compound in Abbotabad. Navy SEALS shot bin Laden dead in a raid on 2 May 2011.

Just one other, less serious example must serve as an illustration. In July 2001, after four years on the run, the authorities finally caught up in the Philippines with Alfred Sirven, a French businessman wanted as part of the investigations into the Elf Aquitaine corruption scandal. At the moment of arrest, Sirven realised the potentially incriminating evidence he held on his person. James Tasoc, of the National Bureau of Investigation in the Philippines, described what happened next: 'He munched up the chip of his mobile like chewing gum. He broke it with his teeth.'

While the uses of mobile phones for criminal activity and detection are revealing, we should not forget a much more everyday role: many mobile phones are carried as part of the complex strategies that we all develop of dealing with the risks and dangers of modern life. So a phone is bought by parents for a son or daughter who is leaving home as a student, or for a teenager who is starting to stay out late at night, as an act of reassurance: should they ever be in trouble, they will not lack the means of contact. The diminishing guardianship of family life is replaced by the constant touch of the mobile phone. Again, the fear of car breakdowns in remote places has often been cited by women as a major reason for owning a cellphone. (Notice how the two technologies of mobility interact again.) Likewise, in countries such as Sweden, the United Kingdom or Germany, where most of the population already own a mobile phone, further expansion of the market has depended on sales pitches that play on fear. Early adopters of the mobile needed little persuasion, and the young took to the chatty communicability of mobile phones like ducks to water. But selling to an older market, which can be either more technophobic or wiser in their purchases, has relied on presenting the mobile phone as a safety technology of last resort. This was why my parents bought one. While anyone who has been saved a walk along the hard shoulder of a motorway at night to find a landline telephone will know the true benefits of the mobile phone as a tool of safety, sometimes the device seems to be merely talismanic. Like a statuette of St Christopher, possession alone seems to ward off danger. A colleague of mine tells a story of one of his parents, who carried a mobile phone in their car. On one occasion a passenger needed to make a phone call and asked to use the mobile. Unsure about the method of operation, the passenger enquired how to work it. 'Oh, I don't know,' came the reply. 'I always have it turned off. It's only for emergencies.'

Women were targeted in mobile phone advertising in the 1980s. Note how 'peace of mind' is promised if 'you're always in touch'. However, exploiting fear of crime was an old tactic, as the next illustration demonstrates. (BT Archives)

'Why aren't YOU on the phone?', issued by the telephone development association, c. 1905 – evidence that fear of crime has been used to sell telephones since the early days. Thank heavens that 'obliging constables' would save us from 'marauding humanity'! (BT Archives)

Chapter 20
Phone hacking:
a very British scandal

The full extent of phone hacking in Britain only became apparent in 2011 when a full-blown controversy over illegal practices broke out. What was revealed was a queasy and disturbing set of relationships between newspapers, celebrities, politicians and police. It started, as with the earlier case of Princess Diana's 'Squidgy' tapes, with royalty. In November 2005, the *News of the World*, a Sunday tabloid which was part of Rupert Murdoch's News International empire, the biggest-selling newspaper in the country and jocularly known as *News of the Screws* for its coverage of salacious gossip, published a story on Prince William's injured knee. Buckingham Palace suspected phone hacking as the source of the story and called in the police. In 2006 the newspaper's royal editor, Clive Goodman, and its private investigator Glenn Mulcaire, suspected of doing the dirty work, were arrested and charged. For many months the newspaper held the line that this was a limited case of rogue reporting. A perfunctory police investigation seemed to confirm this account. In the meantime Andy Coulson, the then editor of the *News of the World* was appointed as the new prime minister David Cameron's director of communications and planning.

However, by 2010 it was clear that many celebrities, national figures and others had had their phones hacked in the early- to mid-2000s. In the analogue days of the Squidgy tapes this would have been done by simply eavesdropping using radio scanners. In the second-generation, digital era of mobile phones the hacking was typically achieved by targeting voicemail, exploiting the fact that many people did not change the factory settings of PINs ('1234', for example). A complementary technique was known as 'double whacking', in which one person engaged a phone number while another person rang in; the second caller would be directed to the voicemail account, which could then be hacked. Mulcaire specialised in the use of another method: he called phone service providers using a unique retrieval number and then used a new PIN.

A small industry of private investigators grew up from the late 1990s, hacking voicemails and selling the information on to willing journalists. One journalist, James Hipwell, who worked the City Desk, located next to the showbiz journalists, on the *Daily Mirror*, described phone hacking as a 'bog-standard journalistic tool for gathering information'. Dominic Mohan, showbusiness editor of the *Sun*, even jokily thanked Vodafone's 'lack of security' during a speech to other members of the press. Nevertheless this was insider knowledge that stood in stark contrast to public denial that the practice was widespread, or anything more than a 'rogue reporter'. Despite repeated attempts to bury the story, in mid-2011, parliamentary questioning – led by Labour MP Tom Watson and investigative journalism by the *Guardian* and the *New York Times* – uncovered the shocking extent of the misdeeds. Hacking was one of a number of so-called 'black arts', that also extended to such practices as 'blagging' (the impersonation of others in order to secure personal information), bribery and harassment. The suspicions hinted at collusion at high levels between senior News International executives, journalists and editors, private investigators and senior police officers. The tipping point came in July 2011 when it was claimed that the murdered schoolgirl Milly Dowler's phone had been hacked – and, it seemed, voicemails deleted, offering cruel and false hope to her parents. Coulson and Rebekah Brooks, chief executive of News International and former editor of the *News of the World*, were forced to resign. The *News of the World*, with advertisers deserting, folded. The last edition, on 7 July 2011, confessed in an editorial 'Quite simply, we lost our way ... Phones were hacked, and for that this newspaper is truly sorry.'

The phone-hacking scandal prompted a wide-ranging inquiry into press ethics, chaired by Lord Leveson, which began in November 2011. A host of witnesses were brought forward and their testimony recorded and probed. The public heard 'how the details of private lives, known only to the witnesses testifying (in other words, the targets of voicemail hacking) and their most trusted confidants and friends, became the subject of articles in the press', while through the technique of voicemail hacking 'journalists and press photographers were able to record moments that were intensely private, such as relationship breakdown, or family grief'. Sienna Miller, star of *Layer Cake* and *Factory Girl*,

> explained [to the Leveson Inquiry] how she was the subject of many articles either speculating on or reporting the state of her relationship with the actor Jude Law. In many cases, the information that had formed the basis of these articles had been known only to Ms Miller, Mr Law and a very small number of confidants who had not shared the information further. Ms Miller gave a graphic description of the fall-out from the voicemail hacking which News International has, of course, admitted took place. This included the corrosive loss of trust in aspects of family life, in relationships and in friendships, Ms Miller assuming, understandably, that her inner circle was the source of stories in the press. She described herself as 'torn between feeling completely paranoid that either someone close to [her] (a trusted family member or friend) was selling this information to the media or that someone was somehow hacking [her] telephone.'

These expressions of feelings of violation, anxiety and intrusion were echoed in other testimony, not only from celebrities:

> Although the targets included a large number of celebrities, sports stars and people in positions of responsibility, they also included many other ordinary individuals who happened to know a celebrity or sports star, or happened to be employed by them. Other victims had no association with anyone in the public eye at all, but were, like the Dowlers, in the wrong place at the wrong time.

Leveson finally reported his findings and recommendations in November 2012. Even then Leveson's weighty tome – a full 1,987 pages – drew back from full disclosure of many of the cases of alleged phone interception in order not to prejudice ongoing criminal proceedings. While Leveson noted that 'it is still not clear just how widespread the practice of phone hacking was, or the extent to which it may have extended beyond one title; and, in the light of the limitations which necessarily impact on this aspect of the Inquiry because of the ongoing investigation and impending prosecutions, it is simply not possible to be definitive', nevertheless the evidence presented 'points to phone hacking being a common and known practice at the NoTW and elsewhere'. On the specific case of the Milly Dowler phone the facts remained murky. The phone had certainly been hacked and presumably voicemails listened to. But the deletions, which raised false hopes and were the direct cause of the spike in public revulsion of the practice of phone hacking, may have been caused by something as simple as a technical system update. Nevertheless, on the behaviour of the *News of the World*, Leveson spoke brutally:

> In truth, at no stage did anybody drill down into the facts to answer the myriad of questions that could have been asked and which could be encompassed by the all-embracing question 'What the hell was going on?' These questions included what Mr Mulcaire had been doing for such rewards and for whom; what oversight had been exercised in relation to the use of his services; why had Mr Goodman felt it justifiable to involve himself in phone hacking; why had he argued that he should be able to return to employment and why was he being (or why had [he] been) paid off. On any showing, these questions were there to be asked and simple denials should not have been considered sufficient. This suggests a cover-up by somebody and at more than one level. Although this conclusion might be parsimonious, it is more than sufficient to throw clear light on the culture, practices and ethics existing

and operating at the *News of the World* at the material time.

Throughout the same period and, as far as is known, entirely unconnected with the phone-hacking scandal, the surveillance of mobile phone records by the British secret state had increased; this took place legally and, one hopes, under ultimate political control. Occasional glimpses can be made of this shadowy world. For example, in 2008 it was revealed that the Government Communications Headquarters (GCHQ, the United Kingdom's electronic intelligence agency) recorded mobile phone exchanges between the Omagh bombers, responsible for a devastating explosion in a Northern Irish town, in 1998. In the 2000s there existed a voluntary agreement that mobile companies would store details of times, dates, duration and location of calls, as well as websites visited and email addresses used, for twelve months. The data were made available to police and security services. Indeed the then Home Secretary, Jacqui Smith, claimed that 'data about calls … is used as important evidence in 95 per cent of serious crime cases and almost all security service operations since 2004'. From 2008 there have been attempts to push legislation through Parliament that would make such surveillance compulsory and extend its reach.

Of course, if the bulk of the British population – celebs as well as civilians – had not been in the habit of keeping in constant touch then surveillance, either by opportunistic crooks or by authorised guardians of collective security, would not be worthwhile. The ubiquity of mobile phones was the first condition that made the phone-hacking scandal possible. But a second point relates to our intimate relationship with this personal technology. We use cellphones as if in a private bubble, even when in public. It is a double shock when our calls, made in private spaces, are eavesdropped. 'While phone hacking itself is a "silent crime" inasmuch as the victim will usually be unaware of, or not even suspect, the covert assault on his or her privacy', noted Leveson, 'its consequences – both direct and indirect – have often been serious and wide-ranging'. The force of this assault is directly in proportion to our mistaken sense of privacy.

Chapter 21
Phones on film

Our sense of danger is powerfully shaped by mediated representations of life. Hitch-hiking is now rare in the West not because hitch-hikers are a danger to drivers, or vice versa, but because on film a hitch-hiker tends to be a homicidal maniac. Paedophiles are a menace, but a youngster is far more likely to be run over by someone driving their kids to school – which they do because 'the streets are dangerous' – than to be abducted. Again a major cause of the school run is our calculation of risk, in this case to children, based on what we read and see, rather than on what statistics might show.

So how we use one technology of mobility – the car – depends on how it has been portrayed on TV, in film or in print. What, then, does the mobile in the media show us? In one straightforward way, the mobile has been a great boon to scriptwriters. A good plot depends on interaction between characters. Communication technologies allow characters who are not in the same room as each other to interact, thereby literally expanding the dramatic range. With a landline telephone, a character is rooted to the spot. But with first cordless phones (which I haven't discussed here) and then mobile phones, characters were set free. The scriptwriters of soaps and sitcoms were perhaps one of the greatest beneficiaries, since soaps and sitcoms depend more than any other televisual genre on conversation and gossip. So an episode of *The Simpsons* or *One Foot in the Grave*, to take just two examples from the United States and the United Kingdom respectively, can even be dominated by characters on mobile phones.

Film both created and exploited mobile phones as iconic markers of status. So Gordon Gekko, the feral corporate raider played by Michael Douglas in the sharp satire of 1980s excess *Wall Street* (1987), barked orders down the brick-like cellphone while walking on an Atlantic beach. In Gekko's individualistic, greed-driven world, money never slept – and never stopped moving. Gekko would have no 'dead time'. *Wall Street* repeatedly drew a contrast between this new atomised anti-society and the older traditional society where personal integrity was based on good character, and a firm's worth lay in real products, not junk finance. The cellphone was the icon of the new. And fiction in turn shaped fact. Across the ocean, the city workers in London who ostentatiously waved mobile phones modelled themselves on Gekko, their hero.

So phones on film could symbolise connection between characters, advertise social status or even stand for the absence of society. In the Nollywood flick *Phone Swap* (2012) the reliance on, and new ubiquity of, mobile phones, in this case in Nigeria, is the source of gentle social satire (as well as a promotion of BlackBerry, one of the funders of the film). Phones can also be, simply, triggers. In *The Hurt Locker* (2008), Kathryn Bigelow's film set among US Army bomb disposal teams in Iraq, they are used to detonate explosives, as they were in real life. But a more intriguing use of the mobile exploited the contradictions – and horror – of being in constant touch. In David Lynch's *Lost Highway* (1997) a creepy atmosphere had already been built up with the delivery of video tapes made by an intruder into the house of characters played by Patricia Arquette and Bill Pullman. Video tapes are information technologies that shift time: the footage merely showed that the intruder had been in the house some time in the past. But creepiness turns to full horror at a party, where the uncomfortable Pullman is approached by a white-faced man with more than a whiff of sulphur about him. The white-faced man insists that they have met. What's more, he insists that he is in Pullman's home at the same instant as he is there at the party. The proof is a mobile phone call. The mobile is an information technology of instantaneous time – the horror comes from the sudden realisation that the white-faced man is in two places at once, so something is seriously wrong.

There are two ways in which mobiles feature in stories of the uncanny, and both reflect contradictory aspects of mobile culture. In series such as *The X-Files*, mobile phones are part of the armoury of good – in this case the FBI agents Fox Mulder and Dana Scully

– against evil. The cellphone here provides horizontal communication between the heroes (Mulder, Scully) who are working outside – and often against – the centralised hierarchical organisation (the FBI).

The same analysis fits *The Matrix* (1999), in which Keanu Reeves and allies discover that the world is a simulation created by an authoritarian mechanical regime. They can act in both worlds only because they possess state-of-the-art mobiles. And as the entertainment and communications industries converged in the late 1990s, *The Matrix* acted as an advert for the particular phone used, the Nokia 8110i. So, ironically, the mechanical simulation-creating regime won:

> Nokia's mobile phones create the vital link between the dream world and the reality in *The Matrix*. The heroes of the movie could not do their job and save the world without the seamless connectivity provided by Nokia's mobile phones. Even though our everyday tasks and duties may be less important than those of the heroes of *The Matrix*, today we can all appreciate the new dimension of life enabled by mobile telephony. As the leading brand in mobile communications, Nokia is proud to see that the makers of *The Matrix* have chosen Nokia's mobile phones to be used in their film.

So said Heikki Norta, General Manager, Marketing

Services, Nokia Mobile Phones, Europe and Africa, on the day of the film's launch in 1999.

In the second way, the creepiness of instantaneous remote communication is exploited. In the *Scream* movies, which knowingly cherry-picked the whole horror genre, the killer was anonymous, remote but also scarily present as soon as the call was made. In *Lost Highway* the uncanniness stemmed from the impossibility of being in two places at the same time – a short-circuiting of spatial logic. How can someone be both present and not present? Mobile phones give us a powerful sense of co-presence that can be shockingly undermined. I was once chatting to a friend who was walking down the Hackney Road in London. There was a scuffle and then silence. While it was clear what had happened – the phone had been snatched – the shock (for me) lay in sudden helplessness: the realisation that someone who seemed near is in fact far. Constant touch is illusory.

The mobile phone on film can counter and cause the uncanny. These two modes are part of longer traditions in storytelling. The use of the latest communication technologies to counter ancient evil is nowhere better illustrated than in Bram Stoker's *Dracula* (1897), in which otherwise utterly ordinary modern Europeans can defeat the Count because they possess Dictaphones, telegraphs and an efficient postal service. (Recall that Stoker's novel, for good reason, is in epistolary form.) *The Blair Witch Project* (1999) was deliberately set a few years in the past, after cheap video cameras (on which the film's beguiling realism depends) but *before* mobile phones. There was no escape for this second team of ordinary humans.

Jeffrey Sconce, in *Haunted Media*, has traced how communications technologies have persistently been associated with the uncanny, from the 'spiritual telegraph' of the 1840s to oppressive other-worlds of fictional cyberspaces in the late 20th century. So with early radio, catching distant voices by accident across the ether ('distant signal' or 'DX fishing') suggested to many authors a metaphor for the fragile bonds between individuals and the potential for traumatic disconnection. 'Stage and screen at the beginning of the century', notes Sconce, 'saw a number of productions that featured distraught husbands listening helplessly on the phone as intruders in the home attacked their families.' Later, episodes of *The Twilight Zone* featured uncanny communication from the dead by phone: in 'A Long Distance Call' (1961) a recently deceased grandmother called her grandson on a toy telephone, and in 'Night Call' (1964) a long-dead man contacts his fiancée via a telephone wire that has fallen on his grave. Horror stems from interrupted mobility: whether it be from confinement to a grave or, traumatically outside fiction, from mobile phones in the twisted wreckage of train crashes, or the last conversations on the hijacked airliners of 11 September 2001.

Part four
Smartphones

Chapter 22
Intimately personal computers

Right at the beginning of this book I said that we should pay attention to the technologies that we carry around with us because they tell us a lot about what we value and why. Very few technologies have made the leap, and each one is important. We wear a wristwatch because we live in a society that is choreographed by reference to a common standard of timekeeping. We carry a comb if we care about what our personal appearance says about us to others. We carried a first- or second-generation mobile phone because communication was desirable, even essential, on the move.

But we are now carrying around a new object, one that might trick us into thinking that it is merely an extended phone, but is in fact, I think, a radically new personal device. The smartphone, towards which the first 3G phones were inching, is not just a phone. It's a computer. And computers are unique – they are, in a crucial respect, unlike any other technology – and uniquely important in the history of the modern world.

Most devices are special-purpose machines. A comb straightens hair. A watch tells the time. A lawnmower, to give another example, mows lawns. The whole design is made to achieve this purpose. While you can use a lawnmower for other tasks – for example, propping open a door – these are limited. A computer, on the other hand, is a general-purpose machine. And it only needs three components to work: a set of instructions, a memory and a processing unit where the instructions work on the data held in the memory. You can make a computer out of almost any material. Charles Babbage designed a machine made of brass, wood and card that was the 19th-century equivalent of the computer. But we use electronics, because electrons can move at unimaginably faster speeds. The first electronic stored-program computers were built in the middle of the 20th century, at a time when the radio industry had spurred the development of lots of electronic components, such as vacuum tubes, and global conflicts increased the demand for fast calculation. These first computers were enormous, filling rooms with racks of valves and wires. But over the following decades the components got smaller and the computer shrank. The new transistors of the 1950s were tinier than vacuum tubes. Wafers of silicon the size of a fingernail, etched in the 1960s, contained hundreds and then thousands of transistors. Smaller devices were more widely useful, creating a powerful feedback effect driving miniaturisation and the spread of computers. In the 1970s small and personal computers appeared. By the 1980s they populated homes and businesses across the world.

We use the desktop computer, the general-purpose

machine, for all kinds of tasks. I'm sitting at one now, writing this text. I'm also keeping an ear out for the live cricket tournament under way in Sri Lanka. I have Skype open in case someone calls, and my music is a click away too. I have Excel spreadsheets open for students' marks, and both Firefox and Internet Explorer browsers running, some linking in to organisational databases, some displaying email, while another is playing a David Foster Wallace inauguration speech on YouTube. Frankly it's amazing I can concentrate. But there are two points to making this list. One, the computer is incredibly flexible for a single machine. And second, it's so useful, so involving, that I'd like to carry these capacities around with me.

There are three concentric rings of 'personal' technologies. The outer ring consists of 'owned' technologies that are mine, that I use but do not move around with me. The desktop computer is an example. It's too heavy, and anyway it is anchored via a nasty mess of wires and cables. In the middle ring are 'portable' technologies. I have a laptop computer which I carry around if I need it. But it's a bit of hassle, and I'm always aware that I'm burdened. If I stop I'll put it down. I'd rather leave it at home. Nevertheless, it is designed to be a 'personal' technology that works on the go – it has a battery that lasts long enough to be able to be useful, and the case has a handle, among other features. Finally, there is the inner ring of 'intimate' technologies. These are portable but are carried without exertion. They are kept close to the body. They are so useful or important or engaging that we don't register their weight. Very few technologies make it through to the inner ring, and some of those that do date from the earliest periods of human existence. Right now my intimate technologies are clothes (Palaeolithic), shoes (ditto), glasses (a medieval innovation) and – my intimate general-purpose computer, my little chip of modernity – a smartphone, an iPhone 4.

The things I use my smartphone for are just as diverse, perhaps more so, as they were in the case of my desktop computer. I check my mail, send texts, update a Facebook status and skim through tweets. I browse, using Safari, websites that carry everything from national news and sport to London ornithology. I share my pictures on Flickr, listen to music on the iPod and play Angry Birds, Blitz and Welder. Under dark clear skies I can hold the iPhone up and read the stars using Starmap Pro. When bored I flick left and right through my apps to find something to do. I even, occasionally, use it to make a phone call.

I was in 'constant touch' with my old Nokia phone in the sense that wherever I was I could talk to my friends, relatives and colleagues. But the 'constant touch' of the iPhone is something more: it absorbs my attention and even when it doesn't I find that I unconsciously reach for the familiar smooth weight. My fingers, eyes and mind are absorbed. And I am not alone – I have been in full train carriages where every passenger was communing with his or her smartphone. Each in a private bubble of constant touch.

Three things came together to help smartphones dominate in this way. None was inevitable. First, the mobile networks had to have the capacity to handle greater amounts of data. This transition was anticipated and pushed in the allocation of and payment for third generation ('3G') licences in the early- to mid-2000s. Second, a device had to be designed that made use of these capacities. As is often the case in the years when a technology is young, there was a huge number of smartphone designs pitched, rejected or launched. Some of them, for example the very first iPhone, depended on older GSM technologies rather than 3G. Nor was the smartphone the first intimate personal computer (handheld 'organisers' date back to the 1980s). However, in retrospect, it is the iPhone that stands out as the device in which key features were decisively introduced. Smartphones were not invented by Apple, but they were defined by Apple. Furthermore, the iPhone did not succeed through the talents of its designers alone. If anything, the closed world of the 'cult of Mac' was a hindrance to be overcome rather than an advantage, at least if the iPhone was to be a truly mass-market device. Thirdly and finally, then, the smartphone had to be discovered by its users, who experimented and found out what the smartphone was good for. Only mobile users can build a mobile culture. Let us look in more detail at each of these three secrets to the success of the smartphone.

Chapter 23

3G: a cellular world made by standards

In 1950, the major ports of the world swarmed with human activity. The job of a stevedore or longshoreman, someone who loaded and unloaded ships, was a skilled one: goods could come in a wide variety of shapes and sizes, and the quickest, most efficient way of moving them had to be worked out. Once dockside, the goods might wait for some time, each minute costing money, before they could be moved to market. Each port had its own system, its own traditions and its own considerable pool of labour, amounting to thousands of dockers. In the 1960s and 1970s this picture of the working dock was transformed by containerisation: goods would be packed in identical steel boxes, making the jobs of lifting on and off ship, and of transporting to and from a port, much simpler (and cheaper). Some historians credit the innovation to the experiments of the United States military in the Second World War, in which essential supplies had to be shipped to Europe and across the Pacific in a form that was secure. Others highlight the entrepreneurial spirit of Malcolm McLean, an ex-trucker whose SeaLand Inc. company began shipping goods in containers along the east coast of the USA from 1956. By the 1970s it was clear that a revolution in the global transport of *material* goods had happened, and that it was underpinned by two fundamental developments.

First, a global technological system for transport went hand in hand with the spread of a single standard. While many agreed that a standard container was a good thing, there had been much debate about what the standard should be. The outcome was a container eight feet high by eight feet wide, with lengths either 20, 35 or 40 feet. (Universal agreement on the width and the height were crucial for stacking containers, the length not so – think about how a sound brick wall can be made with short bricks and long bricks.) If there had been many competing standards in use, then global trade, and with it the forces of globalisation, would have been significantly reduced. Second, as the standard spread – which it did by a combination of commercial and governmental decisions – the old practices and infrastructure had to be torn up and replaced. Starting with Port Elizabeth, New Jersey, container ports were built to move the standardised containers onto ships converted for the new boxes or onto trains and trucks. Some old ports – not least London, which had once been the busiest in the world – died. Others, such as Long Beach, California, were refitted at great cost. The new infrastructure was massively expensive, but was paid for by the savings in transporting goods and in savings of scale: all the world was using one standard. Ninety per cent of the world's trade moves in containers.

A fixed infrastructure and standards mutually agreed beforehand facilitated global mobility: in the case of time zones, mobility of pocket watches; in the case of containerisation, mobility of material goods; and in the case of the internet (where the fixed infrastructure was landlines and the standards were TCP/IP protocols), mobility of *non-material* goods. (There are profound reasons why the latter two cases of means of moving packets, material and non-material, appeared at the same time.) And from fairly early in its history, there have been visions of how a fixed worldwide infrastructure of cellular phones would enable the global mobility of communication. We have seen how very different *national* systems of mobile telephony were built, and now we will see what makes a *global* system and why.

Jorma Niemienen, then president of Mobira, imagined in 1982 that the mobile world could be built on Nordic lines:

> NMT [Nordic Mobile Telephone] is an example of the direction which must be taken. The ultimate objective must be a world-wide system that permits indefinite communication of mobile people with each other, irrespective of location.

In Vancouver in 1986, before the first call had ever been made on the second – i.e. digital – generation of phones, a gathering of telecommunications planners launched the third. Initially called 'Future Public Land Mobile Telephone System' or FPLMTS – 'unpronounceable in any language' writes Garrard, correctly – 'the initial concept for the third generation was very simple: a pocket-sized mobile telephone that could be used anywhere in the world.' Third-generation (3G) mobile phones started as a geographical idea, but as the internet rocketed in the 1990s it provided proof that there was public interest in, and a nascent mass market for, mobile online services. 3G became more and more a plan for mobile phones that would handle data – internet-type services, videos, games – as well as voice. The relative successes and failures, respectively, of i-mode in Japan and WAP in Europe and the United States, were dressed up as rehearsals, generation two and a half, for the data-rich 3G.

Despite the lessons taught by the cases of competing mobile standards in the United States (or indeed, the success of single standards such as Europe's GSM or McLean's containerisation), third-generation mobile has splintered into several different standards. So much was at stake – perhaps the biggest telecommunications sector of the 21st century – that uniform agreement was perhaps impossible to achieve in the face of divergent commercial interests. So International Mobile

Telecommunications 2000 (IMT-2000, the more friendly name for FPLMTS), became the umbrella for five different standards, employing variations of all three means of packaging up and sending data over mobile networks: TDMA, FDMA and CDMA. Each had different coalitions of backers, reflecting the state of a mobile industry that had already become internationalised after a series of mergers and new operations, led by voracious companies such as Vodafone and Hutchison, and by the American 'baby Bell' companies' attempts to expand away from the restrictive home markets in the 1990s. Despite the internationalisation of the mobile sector, an interesting pattern was apparent by 2002: American- and Japanese-based companies were doing better at pushing 3G than their European competitors: in a reversal of the transition from first- (analogue) to second- (digital) generation cellphones, when the United States lost the lead partly because its first generation was too successful, the success of European second-generation systems, particularly GSM, had led to apathy towards 3G.

The first licences for the spectrum space allocated to 3G mobile phones came up for grabs at the very end of the last century. With internet stocks still riding high, and where an auction format was chosen by governments, mobile companies bid against each other, driving the price for spectrum to stratospheric levels. When the bidding was over in the United Kingdom, the government received a windfall of £22.47 billion ($35.4 billion), which was prudently earmarked by the chancellor of the exchequer, Gordon Brown, for paying off part of the national debt. Licences went to the four existing operators – Vodafone Airtouch, One 2 One, BT Cellnet and Orange – and a newcomer: a conglomerate, backed by the Hong Kong-based Hutchison Whampoa, which launched its service called, simply, '3'. In Germany, a bigger potential market than Britain, the auction raised $45.6 billion, five times the amount initially expected. France refused to run such a market-driven scheme and preferred to retain central control, offering four licences at fixed prices of $4.6 billion each. Only two were taken up. Sweden, even more planning-minded, awarded licences at $10,000 each, plus a cut of profits. The German and UK windfalls were watched jealously in the United States, where the practice of local auctions was again followed for the sale of broadband Personal Communication Service (PCS) licences, with the outcome in early 2001 that a mere $16.86 billion was raised for 422 licences (113 of which went to a joint venture between Vodafone and Verizon Wireless).

But what had caused this American shortfall? By the time of the US auction the internet stock bubble was bursting, and the giant bids that had seemed necessary to secure important territories and markets in earlier months now seemed decidedly dicey. Indeed, the bids made by companies such as Vodafone were justified by appeal to the strength of their stock market value, and as these slid alongside other telecoms stocks, the expenditure looked more and more untenable. Moreover, 3G could not operate on the existing infrastructure. Entirely new networks of base stations and mobile switching centres needed to be built. Like the containerisation of ports, the great capital outlay in a gamble on a new standard was to be new infrastructure (the costs are comparable). By 2002, the licences were active but, apart from in places such as the Isle of Man, a third-generation experimental island, very few services were launched until 2003. The success of third-generation mobile phones, at this point, depended on the unknowable willingness of the public to buy them, and without good content – in the form of addictive entertainment or really useful services – a repetition of the WAP debacle was possible. On the other hand, there were great hopes that the third generation might prove to be like a global i-mode, to the great relief of the world economy.

As we have seen, sales of licences for third-generation (3G) networks raised immense revenues for European governments, less so in the United States where the auction was held after the crash in the technology stock market. Similar networks were launched in Japan, but elsewhere in 2002 you had to be in odd places like the Isle of Man to see what all the fuss was about. In the United States, 3G has followed the patchwork pattern that had been established with earlier generations of cellphones. So, for example, as early as December 2001, Verizon offered a 3G service – so long as you happened to live in a corridor of land from Norfolk, Virginia to Portland, Maine, or in the Salt Lake City area, or around San Francisco and Silicon Valley. (What is more, it would only work if you plugged it into a computer, so was not very convenient.) Further 3G services were launched soon after in cities such as Chicago and New York. Sprint, a company that has always concentrated on long-distance calls, souped up its old network to carry 3G nationwide in August 2002. Fanfares also greeted the launch of 3G in Europe, including Finland and Austria (September 2002), parts of Russia (October 2002), the United Kingdom (March 2003), Italy (May 2003) and Slovenia (December 2003).

But it quickly became apparent that launching 3G was quite different from turning on an old-style mobile network. With an old cellphone system, once it was turned on you could make calls, and that was pretty much it. But 3G promised a cornucopia of data services, and these weren't all ready at once. The bounty remained firmly fixed in the future. (Imagine! It will be just like sitting at your computer: you'll be able to do anything that you could via the web, except that you'll have to squint.) So, in the early 2000s 3G crept out, with a dash of video messaging here and a smattering of live football and games there. It would be half a decade before we experienced what 3G really could do.

The launch of 3G witnessed a curious, reflexive twist on a familiar pattern in the history of communication technologies: the ways that the typical real-world 'use' of a technology has often been one discovered by users rather than that anticipated by producers. So, for example, Marconi thought radio would be primarily a means of sending code not voice (indeed he called it 'wireless telegraphy'). Producers of early telephones sold them as one-way ordering devices, not instruments of trivial two-way chatter. Both email and text messaging were afterthoughts, mere secondary applications in the eyes of the designers of packet-switching and digital cellular networks. We have also seen how M-PESA as primarily a peer-to-peer money transfer system was the discovery of Kenyan mobile users rather than the anticipated main purpose of the scheme's architects. In all five of these cases it was customers who pioneered the typical pattern of use. As 3G-type services like video messaging were added to phones, it seemed that for the first time in the history of communication technologies, the producers were not only aware of this pattern, but were also banking on it happening again. So in late 2002 – before 3G proper – Vodafone deployed the Manchester band The Mock Turtles to ask their UK customers: 'Can you dig it?' Quite a lot of money was riding on the answer to this question. If the users did not 'dig'

3G by finding its typical use, soon and with gusto, then billions of pounds and dollars, euros and yen would be lost. In fact, while video messaging failed (as it always does), the smartphone, and with it 3G, has thrived.

By 2005 some of the 3G hopes were being realised. Old players such as Vodafone, as well as new such as Hutchison Whampoa's 3, were attracting custom as coverage improved and new handsets launched. 3's new X-series of mobile phones, for example, offered a phone that carried Google searches, access to the online auction site eBay, communication via Skype, and even a television service. Furthermore, in a move which reassured customers, a flat rate was introduced for data downloads. (Nevertheless, a frequent news story in the mid-2000s featured panicked customers, often parents who had lent their mobiles to their offspring, only belatedly finding out just how expensive an unre-strained bout of data consumption could be.) Even so, the enormous outlay on 3G licences meant there was nervousness, especially because the mobile technolo-gies were changing so fast. One example of this change was the rise of Wi-Fi hotspots in the same period. A Skype video call over a Wi-Fi network saved money for the consumer but had to be written off as lost income to the network operator. But an even greater change was the entry of a new player to the mobile market, one which would redefine what a mobile could be: Apple's smartphone.

Chapter 24
Apple

In 2003, an innovative company launched a smartphone that would prove to be a great hit with consumers, who so loved the look and feel of the device that they became engrossed in their own world with it. Soon the company was riding high, with more and more customers signing up every day. But it was not to last.

The company was Research in Motion and the product was the BlackBerry. The Canadian firm had built two-way pagers in the late 1990s, a modest hit. But their BlackBerry smartphone of 2003 proved to be the breakthrough. It was a handheld, intimate phone with a basic black-and-white screen. It was the size and weight of a pack of cards, held horizontally. So far, nothing unusual. But the BlackBerry also had a well-designed miniature QWERTY keyboard, over which thumbs could rapidly and easily flick and press. When combined with its other outstanding feature, a secure channel for email, the BlackBerry provided an extraordinarily powerful device for keeping in constant touch with work. BlackBerrys were very popular with big organisations, and many government and corporate employees received them as part of the job, along with the expectation, as with early analogue cellphones, that the 'dead time' of travel and leisure would now be spent to the employers' benefit. This expectation was met. The ambitious saw the BlackBerry as a tool to work hard. The device had to be checked every spare minute just in case a key office manoeuvre or managerial decision was missed. Never turned off and always to hand, intensifying the trend of transferring work into the home, it is no wonder that the device was dubbed the 'CrackBerry'.

The BlackBerry was very successful in its niche as a personal portable email browser and telephone. It also became curiously popular with teenagers, who were attracted by a free direct messaging application that came rolled into the device. But there were plenty of other smartphones around in the mid-2000s, as a flick through a contemporary Carphone Warehouse or Phones4U sales brochure proves. There are products from Motorola, Nokia, Hewlett-Packard, Sony, Dell and Palm, among others. While there was quite an exciting diversity of design – with flip-up covers or keyboard wings – the overwhelming look was of shiny chrome, plastic casings and tiny buttons crammed around screens. They were chunky and ate the battery before breakfast.

Steve Jobs, the CEO of Apple, loathed them. 'We would sit around talking about how much we hated our phones,' he said, as his biographer Walter Isaacson records. 'They were way too complicated. They had features nobody could figure out, including the address book. It was just Byzantine.' But smartphones, which as intimate personal computers could do so many tasks, were also a threat to Apple. The company had watched as the camera phone had eaten away at the market for digital cameras. Soon smartphones playing digital music might sweep away the mp3 player, including Apple's own iPod.

The way that Apple responded to this threat drew deeply on the values held by the company since its earliest days: a perfectionist focus on well-designed products, a desire to frame the users' experience on Apple's terms, and a closed and controlled approach to innovation. Quite unexpectedly the iPhone, embodying these values, would demonstrate that the 'walled garden' approach to mobile culture and business, which had failed with WAP and never left Japan with i-mode, could succeed. The Apple story is well known, but a brief retelling will remind us how deep and consistent these values have been.

Apple was born in California, for good reasons. The west coast of the United States, and California in particular, had benefited from the industrial expansion of electronics during the Cold War. Missiles, aircraft and big mainframe computers all required smaller and smaller electronic components. The integrated circuits produced by Fairchild Semiconductors were designed in Palo Alto. Nearby, the older firm Hewlett-Packard, started in 1939, had grown to be a big local employer and, crucially, trainer in the skills of electrical engineering. Around town there were plenty of people tinkering with electronics in garages in the evenings. When the first microprocessors – computers on chips – went on sale in the mid-1970s, the Californian tinkerers enthusiastically played with them, exploring what they could do and demonstrating their achievements at venues such as the Homebrew Computer Club. Soon the more skilful amateurs built these chips into rudimentary small computers, and these were even sold in kit form to other enthusiasts. One partnership that formed around this micro business was between Stephen Wozniak and his friend Steve Jobs. Woz was the talented hacker of electronics, capable of sustained bouts of engineering and programming. Jobs was the fixer, securing deals for the chips and caring about the design, packaging, peripherals and sales. Their first Apple was a kit computer, distinctive only in that it could be plugged into a TV, that sold to a handful of people in 1976. Their second, the Apple II, on sale the following year, was the Apple values made manifest: a personal computer with innovative features (such as colour), designed with careful attention not only to outside looks but also to the internal architecture, and sold as a general consumer device. Six million were eventually shipped.

The Macintosh, or Mac, of 1984 continued this pattern of selling slightly high-end, well-designed personal computers to those who saw themselves as creative and discerning. Whereas IBM would license out its schema for a personal computer to any manufacturer, resulting in a flood of cheap and cheerful, but often ugly, desktop machines, the Mac was produced to Apple's design by Apple alone. Whereas the word processing and operating system software for the clones was generally Microsoft by default, due to its dominance in the market, the software on Apple computers was by Apple's invitation only, and usually done in-house. This closed approach meant that software and hardware could be carefully designed together for maximum efficiency. With little waste, Apple could get the most out of the machines, which in turn meant that the Mac could run visually impressive software – not least the icons, mouse and windows interface inspired by Xerox. The lesson was learned: close control over hardware and software led to efficiency of product design which made dramatic new features workable, which in turn attracted custom.

Steve Jobs prized these values. When he was forced out of Apple in 1985 the company drifted and its market share eroded as other companies caught up – not least Microsoft, which offered its own Windows operating system for PCs, replacing its inscrutable text-based MS-DOS. Nevertheless Apple did launch one of the first 'personal digital assistants', the Newton, in 1993. These were limited notepad computers in which a stylus could tap and write on a screen. When Jobs returned in

1996, and then took over as CEO in 1997, he returned the company to the values exemplified by the Apple II and Mac. Fun and funky design, for example, marked out the iMac, a colourful fat wedge-shaped computer launched in 1998. (The name iMac, which would of course be echoed in the iPhone, was one of five monikers pitched by the ad firm TBWA\Chiat\Day. Jobs disliked them all at first.) More significant, though, was the launch of the iPod in 2001. This digital equivalent of the iconic Sony Walkman distinguished itself from the host of mp3 players already available in two ways. First, it was fantastically cleverly designed, thanks to Jonathan Ive: the iPod was an otherworldly thin white block with an intuitively simple 'dial' interface. Second, the iPod worked with iTunes, a piece of software that acted as both online market and personal music library. It locked listeners into a closed world, saved the music industry, which was panicked by illegal downloading and sharing of music files, and made Apple another fortune.

Apple dipped into the mobile phone business tentatively at first. In 2004, Motorola, the granddaddy of handset manufacturers, had launched its Razr, a slim clam-shape phone that sold well. Jobs and the CEO of Motorola, Ed Zander, agreed an unusual collaboration: in 2005 the Rokr phone appeared, essentially a Motorola phone with iTunes preloaded. Users had something like an iPod and cellphone in one. Motorola gained by tapping the popularity of iTunes, while Apple could experiment with the cellphone world in a low-key way.

Somewhere in the secretive back-labs of Apple's Cupertino headquarters, the plans for the iPhone were begun around early 2005. There was clearly a lot of experimentation. This behind-the-scenes activity was only accidentally revealed in 2012 when an Apple vs Samsung patent dispute obliged Apple to describe its process of design. 'Some of the early prototypes of the iPhone are bizarre,' reported Nick Bilton in the *New York Times*:

> One, a long black rectangle, looks as if it is twice the size it should be. Others have beautifully curved glass screens. Another resembles an old silver iPod that just happens to be a phone, too. And there's the strangest of all: an iPhone that looks like a stretched hexagon made of cheap black plastic.

For a while two candidate iPhone designs were seriously considered, one a kind of iPod phone, with a trackwheel, and the other a rectangle with a touch-screen interface. The 'multi-touch' screen was the first of the great innovations introduced by the iPhone. The choice was made by Steve Jobs, who favoured simple, clean design and loathed extraneous buttons or, worse, the stylus. 'God gave us ten styluses,' his biographer reports Jobs repeatedly saying, 'let's not invent another.'

Think about how you use an iPhone. There is one button on the front which simply calls up the main screen. You can swipe to the left, dragging your fingertip across the screen, to move and see other apps. The four apps at the bottom remain the same, and remind us what the main features of these smartphones are: a phone, a music player, a mailbox and an internet browser. If I want to choose to do something, I tap one of the tabs. Sometimes I push with my finger to scroll up or down. Sometimes, and this is very clever and intuitive, I zoom in to something by spreading my fingertips apart. I zoom out by pinching the screen.

Yet when I rush off and throw my phone in my bag, even though it jostles with pens and books the phone does not accidentally ring my aunt or delete my mail. The phone screen is sensitive to live fingers and not inert objects. All this cleverness is possible because the iPhone screen has several smart layers. On the outside is tough glass. (It is called 'gorilla glass', and had been invented – but never developed further – by the Corning company in the 1960s. It is now one of Corning's main products.) Sitting on the glass is a nearly invisible grid of fine electrical wires. You can glimpse this grid if you give the screen a good clean and then peer closely at it under a bright light. The lines are about a millimetre apart. One line carries an electrical charge, while the other detects the slight disturbance caused in the electrical fields as your finger, acting like a weak capacitor, swipes the screen. By arranging the lines in a grid the position of your finger, and whether it's moving in a particular direction, can be measured. The computer in the iPhone then figures out what these prods and swipes might mean, and launches apps or changes the screen accordingly. It's very fast and very smooth, and it means that the whole hardware of traditional phone interfaces – the physical buttons – can be ditched in favour of software and virtual buttons. Apple at first didn't know how to do this. But one company that did was FingerWorks, a spin-off from the University of Michigan. Apple snaffled the company up in a low-profile purchase in 2005, importing the crucial knowledge, skills and patents of the FingerWorks inventors.

The other great innovation of the iPhone was the 'App Store'. Apple's fans had become used to the idea of shopping in a secure online space though iTunes. With applications, or 'apps', users could buy and download pieces of software. In the first iPhones the purchases were made through iTunes, but later models had the App Store as a preinstalled tab. As with iTunes, Apple hosted the market, took a cut of the revenue and acted as a gatekeeper. Any company producing apps would have to secure approval from Apple, which in turn set the bar high, making sure that running the app would be fast, free of bugs and not interfere with the smooth operation of other iPhone programs. Nevertheless, from Apple's point of view this very slight opening up of the iPhone represented a significant risk, and ran against the ingrained values of the company.

Users accepted the restricted choice in exchange for the implied guarantee of quality. In fact they seem to have greatly enjoyed playing in Apple's walled garden. The App Store has been extraordinarily successful: from 500 initial apps in 2008, users in the summer of 2012 had a choice of 700,000. By then over 30 billion apps had been downloaded. Apps come in all different kinds. There are apps to translate languages, design tattoos, teach you to drive or be your personal trainer. Almost all media, from traditional print newspapers to Twitter, have made apps their interface. But the biggest sellers – and the apps that have absorbed most of our time, so far – have been games, such as Rovio's Angry Birds and Halfbrick Studios' Fruit Ninja.

Apple launched the iPhone at the January 2007 Macworld gathering in San Francisco. Steve Jobs, at the top of his game, wowed the home audience. He began speaking at 9.15am, announcing: 'We are going to make some history together today.' For half an hour he cycled through minor achievements and developments, including the underwhelming Apple TV. Then he stepped up a gear. 'This is a day I've been looking forward to for two and a half years,' Jobs began. The audience hushed. He continued: 'Every once in a while

The iPhone.

a revolutionary product comes along that changes everything. One is very fortunate if you get to work on just one of these in your career. Apple has been very fortunate that it's been able to introduce a few of these into the world. In 1984 we introduced the Macintosh. It didn't just change Apple, it changed the whole industry. In 2001 we introduced the first iPod, and it didn't just change the way we all listened to music, it changed the entire music industry.' With such an ancestry, the 'revolutionary product' would have to be extraordinary. But what's this? Jobs:

Well today, we're introducing three revolutionary new products. The first one is a widescreen iPod with touch controls. The second is a revolutionary new mobile phone. And the third is a breakthrough internet communications device.

The crowd had cheered the first, redoubled their noise at the second and scratched their heads at the third.

An iPod, a phone, an internet mobile communicator. An iPod, a phone, an internet mobile communicator ... these are NOT three separate devices! And we are calling it iPhone! Today Apple is going to reinvent the phone. And here it is ...

Eventually, after a litany of complaints about the tiny keyboards and rubbish design of existing smartphones, the simple oblong iPhone appeared on the giant screen behind Jobs, looking like a grey and silver version of the alien monolith from *2001: A Space Odyssey*. The slick demo rushed through the iPhone's features – the multi-touch screen, the 'pinching' of images to zoom in, the camera, the easy access to the web and to phone contacts. 'You can touch your music,' Jobs says. Indeed, touch does everything.

Several versions have followed. After the initial iPhone of 2007, which ran on old GSM standards, came 2008's iPhone 3G and 2009's incrementally improved iPhone 3GS. The squarer and faster iPhone 4 was released in June 2010, again followed by a slightly different iPhone 4S, which included the voice-controlled personal assistant 'Siri'. The taller iPhone 5 arrived in 2012. While all of these new iPhones were launched with characteristic fanfare, and were perfectly delightful products, none of them was as revolutionary as the hype suggested. Certainly, none marked the genuine qualitative leap in terms of design and consumer expectations of the original iPhone.

Steve Jobs at a press conference on the iPhone 4.

However each new tweak of the iPhone served to keep the sales, which started high, extraordinarily profitable, as consumers traded in older versions for the new, fashionable device. Over a million were sold in the first year. In 2010, 40 million iPhones were sold, while the figure for 2012 was well over 100 million. The iPhone has reached all countries. In some, notably China, where China Unicom only signed a deal with Apple in 2009, unlocked iPhones were in widespread use even before they were 'officially' available. The sales were enough to push Apple beyond ExxonMobil and to become, in mid-2012, the world's most valuable company ever. From such a peak the only way is down.

Chapter 25
Apple's rivals

Just as there were smartphones before the iPhone, so, afterwards, Apple's rivals responded with new handsets, many of which were uncannily like the innovative Apple product. Partly this was the result of all the manufacturing companies having available to them comparable advances in technology; partly it was reverse-engineering and imitating features that may or may not be the intellectual property of others. By 2012 many touch-controlled smartphones were being made by large-scale manufacturers, and the arguments about patent infringements had reached the courts.

The most dramatic story was the entry of the Korean firm Samsung. Rather like Nokia, Samsung's origins were as a general-purpose firm, importing and exporting a range of goods, but unlike the Finnish firm it has remained a conglomerate. In the 1970s and 1980s it produced domestic electronic goods for a global market, from microwave ovens and CD players to fridges and televisions. In the 1990s it became a world leader in certain components of electronic devices, not least memory chips and liquid crystal displays. Samsung, even in South Korean terms, has been extraordinarily successful, with a corporate revenue equivalent to the GDP of a moderately wealthy country. Samsung's mobile phones began in 1992, around the time that the second-generation, smaller, digital handsets were being introduced. Nevertheless, South Korea restricted the development of smartphones until relatively late – 2009 – which meant that any South Korean company wishing to sell them had to find ways of learning fast.

Samsung, with its range of manufacturing, could produce many, but not all, of the components needed to form a smartphone. Crucially, the South Korean firm teamed up with Google. The Californian search engine giant, headquartered a stone's throw from Apple, had developed an open operating system for mobile phones called 'Android', based on Linux. (This was despite the fact that Google's Eric Schmidt had sat on the Apple board while the iPhone was under development. Steve Jobs was furious.) Rival operating systems, such as Windows Mobile, Symbian (the market leader, controlled by Nokia), RIM's for BlackBerry, and, of course, Apple's iOS were proprietary and closed. Google's bet was that Android's open nature would make it nimble, quicker to develop new applications for, and appealing to the broader community of coders. Google would still make money because Android users would be led to use its search engines, and the company would accumulate the profit from the ads. Samsung, on the other hand, saw a free tool, one which it did not have to pay a hefty licence fee to use, and one that would be maintained, just as Linux was, by the crowd. (Microsoft, for example, charged between $15 and $25 per handset to use Windows Mobile in 2009. Android would cost nothing.) Many companies came together to promote Android, including other handset manufacturers such as LG (also South Korean), HTC (of Taiwan) and Motorola (of the United States). But it was the alignment of the giants, Samsung and Google, that was most significant. Note, though, that none of these immense corporations had the closed attitudes of Apple. Indeed, even while embracing Android, they continued happily to make phones using other, licensed operating systems, notably Windows Mobile. The overall point is one of contrast: the end-to-end control of Macworld's iPhone vs the shifting corporate alliances and open deals among its rivals.

Samsung's first smartphone using the Android operating system was launched in 2009, and a small army of handsets carrying the name 'Samsung Galaxy' were launched around the world in the following few years. Their diversity is an illustration of the loose-ended, diverse environment in which Samsung operates. They sold incredibly well. But the Galaxy phones had several things in common: touch screens, gesture control and apps available via an app store. Of course these were also things in common with the iPhone.

'Intellectual property' is an elastic term that covers several ways of claiming ownership of an idea. 'Copyright', for example, generally covers text, and is

Samsung's Galaxy phones have fast become rivals to Apple's iPhone.

the legal right of an author to say that they wrote something and the 'copyright' owner alone – with a few exceptions – has the right to allow others to copy the work. I have copyright of the text in this book, and I allow Icon, the publisher, to reproduce it. A 'trademark' is a second type of intellectual property. It is a sign, usually a name or a designed logo, that tells a consumer the source of a product. On the spine of this book is a trademarked logo telling you that this book came from Icon. The third type of intellectual property is the 'patent'. Patents are an exchange – revelation for protection – that was devised, ironically, to prevent secrecy stalling broader economic benefits: the inventor reveals details about a new inventive idea (which then becomes public knowledge) in return for a government-enforced monopoly on its exploitation for a set period of time. While this book can't be patented, almost every device or invention mentioned will have been covered by a thick layer of patents. From its origins in the 19th century, the telecommunications industry has grown up in a tangled forest of intellectual property; fiercely protected and, more often than not, just as fiercely disputed.

As its rivals began to eat away at Apple's smartphone dominance, so the Californian company became ever more litigious, claiming that the similar look and feel was no coincidence but was in fact due to the theft of intellectual property. Soon after Samsung launched its Galaxy phones in 2009, Apple filed a lawsuit in the United States District Court of Northern California, claiming that the South Korean firm had violated patents and trademarks, especially in fashioning the look, design, interface and even packaging of its smartphone. The lawsuits would multiply, not least because intellectual property law varies from region to region. Samsung also responded in kind. By 2011 – the first year in which Americans were buying more Android phones than iPhones – there were well over 50 lawsuits in ten countries. They covered tablet computers as well as smartphones – not surprising, since the former were increasingly mere scaled-up versions of the latter: think of the similarities between the iPad and iPhone. In the courts, home advantage seemed to count: Samsung won most of its case in the South Korean courts, the Californian court sided with Apple in 2012, and the German and British courts fell in between. Furthermore, while Apple vs Samsung was the battle that caught the eye, there was a generalised 'patent war' involving nearly all of the big companies – Microsoft, Motorola Mobility, Google, LG, Sony, HTC, Nokia (which was forced to make Symbian open-source in response to Android), RIM and so on.

Actually, these court cases were the sites of slug-fests between international corporate giants, electronics companies that depended, to a significant degree, on shared ideas and intellectual property. By one estimate, a single smartphone might be entangled in more than 250,000 active patents – not only all the patents relevant to telephony, but also those implicated in computers. What was driving the litigation was not so much outrage over the theft of ideas as competition in an increasingly crowded market by companies whose products were convergently similar.

Chapter 26
Blood on the smartphone

As I am writing this – in October 2012 – the number of smartphones was estimated to have topped 1 billion. (A quarter of this market is in China, with less than one in five in use in the United States.) This 21st-century market is huge and lucrative. With so much at stake, and competition so fierce, manufacturers are pressed to cut costs and drive hard bargains. And behind attractive prices lie exploitation.

Smartphones are made in a globalised, profit-driven world. The design might be made at headquarters, but the contracted manufacturer will be elsewhere, most likely in a part of the world where labour costs are low and conditions are poor. There are other crucial factors too: skills are needed, not least the engineering skills necessary to oversee the rapid and accurate production of a device. Furthermore, manufacturers need to buy in most of the components, which will come from a host of smaller suppliers. There is, therefore, a long supply chain stretching from brand owner and designer, through the principal manufacturer, to smaller companies. Each link is contractual and everyone looks for the deal that keeps costs down and profits up.

Take Apple as an example. Apple's headquarters is in Cupertino, California. The primary manufacturer for

Apple is Foxconn, a Taiwanese firm that runs factories in China. (Apple are not alone here. Foxconn, which started only in 1974, now makes two-fifths of the world's consumer electronics, including the Kindle, the Xbox 360 and the Playstation 3. It may say 'Amazon', 'Sony' or 'Microsoft' on the outside, but Foxconn made it.) Foxconn in turn relies upon a host of smaller suppliers for components. Since 2004, the conditions in which Apple's products were made in Foxconn's Chinese factories have been repeatedly held under close media scrutiny. Allegations include long working hours, grim accommodation, the use of dangerous chemicals, aluminium dust explosions, underage employees and pressure enough to drive workers to suicide.

The *Mail on Sunday* was one of the first to draw attention to the conditions at Foxconn's factory in Shenzhen, China, in 2006. Many others have followed. Personal testimony is difficult to provide, not least because people fear for their livelihoods, but here is how one informant described Foxconn's working conditions to the *Guardian*:

Liu Jingjing, 20 (not her real name), started at Foxconn's Zhengzhou factory six months ago. The work on an iPhone production line is demanding: at first Liu found her fingers and forearms ached every day. After deductions for accommodation and food, she earns just £90 some months. Late last year, as Christmas sales of the latest Apple phone were booming, she says overtime rose, with employees asked to work up to 100 hours a month – despite Apple's 60-hour limit.

She is one of those fortunate enough to be able to sit down on a shop floor where staff usually stand. However, she says employees are only allowed to perch on a third of the seat – to keep them 'nimble'. Liu broke the rules. But because she did not sit in a standardised way she was punished by being asked to write a 'confession letter'. Recently, when she complained about a cut in overtime pay, her boss told her: 'If you do not want to work here, you can get out.'

Chinese workers on the production line at the Foxconn plant at Shenzhen, Guangdong province, 2010. (Press Association)

While the style of manufacturing was shared by other leading companies, there are several reasons why critical media attention focussed on Apple. It was one of the richest companies in the world, making high-end products, and was not shy of portraying itself as wholesomely anti-authoritarian. There was a whiff of hypocrisy. Second, the luxury nature of the goods provided a dramatic contrast to the shoddy conditions being revealed, which made for good copy. Also, Apple was well known. This recognition meant that a debate on globalisation, which was otherwise hard to make concrete, could be brought to life. American political debates on job creation and outsourcing, particularly pertinent during a global recession, formed the context that encouraged the highlighting of Apple's woes.

Apple's first computers, including the bestselling Apple II, Macintosh and even iMac, were designed and assembled in the United States. But by 2004 manufacturing was moving abroad. Chinese factories were cheaper and more flexible. Charles Duhigg and Keith Bradsher, in the *New York Times*, give a particularly telling example. Recall that the iPhone's scratch-free 'gorilla glass' was an American invention, by Corning Inc. However, cutting up the glass and assembling the pieces into iPhones was a different matter. A Chinese factory, underwritten by the Chinese government, bid for the contract and, promising immediate and flexible production, got the job. 'The entire supply chain is in China now,' they record a former Apple executive as saying. 'You need a thousand rubber gaskets? That's the factory next door. You need a million screws? That factory is a block away. You need that screw made a little bit different? It will take three hours.' By 2011, when President Obama and Steve Jobs crossed swords on the issue, Apple employed 43,000 people in the United States, but Apple's contractors elsewhere employed 700,000. 'Those jobs aren't coming back,' said Jobs.

At the beginning of this book I took an old cellphone apart, and I said that global politics could be found in its components. This is every bit as true for smartphones. However, it can be almost impossible to trace the links. Apple, in particular, makes investigation hard. In line with its closed culture of secrecy and control, it is difficult even to lift the lid off the back of the iPhone. (Walter Isaacson, in his biography of Jobs, records that 'when Apple discovered in 2011 that third-party repair shops were opening up the iPhone 4, it replaced the tiny screws with a tamper-resistant Pentalobe screw that was impossible to open with a commercially available screwdriver'.)

But if you wrench the back off and have a look inside, what do you see? The biggest component by far is the lithium ion battery, a big fat thing oddly reminiscent – and this will only make sense if you have gutted a fish – of cod roe. There's a tiny jewel of a camera in the top corner. With the battery and some of the antenna out, we can see the logic board, squeezed and compact. The biggest chip is the processor, made – ironically but significantly – by Samsung, Apple's rival. The central processing unit uses a design licensed from ARM Holdings, a typical British technological success story – a company that does the thinking and licensing but not the making. There are lots of smaller components: chips for GSM frequencies, an accelerometer, a flash memory (also by Samsung). Companies providing parts include Texas Instruments, Broadcom, Infineon and Skyworks Solutions.

Stripping down an iPhone is very much against company wishes. Indeed, Apple only released information identifying its contractors in January 2012 in

An iPhone 4, taken apart. Image courtesy of iFixit (www.ifixit.com).

a step to counter criticism. Its 'Supplier Responsibility Report' makes interesting but incomplete reading. We are told about the 'zero-tolerance for underage labor' and the regular environmental audits. There's even a list of suppliers – 156 of them, from AAC Technologies Holdings Inc. to Zeniya Aluminium Engineering Ltd. But the report doesn't cover the conditions beyond the immediate suppliers, and we don't find out which company supplied what part and how much was paid. By one estimate, 90 per cent of the components in an iPhone were manufactured outside of the United States. Sourced from all over the world, the components are assembled to make an iPhone in China, and the final product is more likely to be bought in China than in any other country.

Chapter 27

Smartphone culture

The phone, like any form of communication, can be analysed in two overlapping ways: by paying attention to its technological structure and to the information that passes through it. The great Canadian commentator Marshall McLuhan called these the medium and the message, and argued that one decisively shaped the other. The old telephone was a classic case of what McLuhan called a 'cool' medium. A tiny channel of communication was opened up when answering a call – a disembodied voice heard over a crackly line. Concentration was needed because most of the cues we need to easily follow a conversation – the movements of hands and head, the eye contact – are missing.

As the phone has passed from landline to mobile cellphone to smartphone, so the channels of communication have become immensely richer. If we wanted to – and largely we don't – we could routinely hold video conversations instead of re-enacting the old telephone style. Actually making calls is not how we have used the new richness of smartphone communication. Rather we have used the widening of the channels – the gush of broadband information – to make the smartphone our primary personal cultural node, our way of recording, transmitting, storing and manipulating meaningful messages. Smartphones are increasingly the tool by which we manage culture.

Let us explore this development by looking at three types of cultural data: images, locations and play. Cameras started appearing in cellphones, as a routine feature, in the very early 2000s. Since then the quality of the lenses, and the size of the images produced (as optimistically measured in megapixels) has increased, to the extent that the digital camera industry, barely coming down after its victory over 35mm plastic analogue film, is in crisis. Camera phones, like the earlier mobile phones, provoked anxieties about privacy, although in an interestingly inverted fashion. Speaking on the phone risked a breach of one's own privacy, and the complaints were of hearing – or rather, overhearing – too much personal information. An example would be on a crowded train in which all but one of the passengers are frozen, embarrassed, as the carriage is filled with the sound of one voice recounting the details of a medical examination. Public space was being mistaken for one person's private space. Camera phones, inversely, were a potential invasion of one's privacy. In both cases the social breaching was most evident, and most discussed, in the early formative years of the technology.

For example, in 2004 the anti-surveillance group Privacy International campaigned against the invasion of the camera phone. 'The misuse of phone cameras', said the director of PI, Simon Davies, was 'becoming a real threat to privacy.' The danger was that people were being photographed without realising it. One of the reasons was that people saw a phone as just that, and not as a potential camera. He called for it to be compulsory for phones to flash every time a picture was taken. Another option, pursued in countries such as South Korea, was to make it impossible for users of camera phones to disable the entirely vestigial 'click' that accompanies the taking of a digital photograph. 'Unless action is taken immediately', said Davies, 'there is a risk that social intimacy will disappear within a decade.'

Social intimacy is not dead yet. One reason is that as camera phones have become commonplace, so societies have improvised, as they do, formal and informal rules for their use. Another reason is that the camera phones are so useful that we have accommodated some of the loss of privacy. Gerard Goggin, the foremost scholar of emerging cellphone culture, cautiously noted in 2006 that the 'social and cultural functions of camera phones are quite distinct' from older camera cultures. In Japan, an early adopter of the camera phone, for example, Fumitoshi Kato and colleagues had noticed that they were used for 'taking photos of serendipitous sightings and moments' rather than the 'special planned events that have traditionally been documented by amateur photography'. And as camera phones became ubiquitous, so any newsworthy event was equally likely to be caught serendipitously. In Britain this phenomenon was dramatically exemplified by the 7 July 2005 bombings. Grainy camera phone footage of passengers picking their way through the smoke and carnage of underground train wrecks defined the initial media coverage. It was taken by the passengers themselves.

In a more mundane sense, of course, camera phones are now an everyday tool for taking, storing and sharing photographic images. A picture at a family celebration will now be most likely taken on a phone rather than an old-style camera, digital or analogue. Any public event, from a concert to a coronation, will be greeted with a forest of hands holding smartphones aloft. As smartphones act as our intimate personal computers, we can use software and websites to manage these pictures. To take one example Flickr, a digital photo-sharing website that began in 2004 and is now owned by Yahoo, has over 50 million members. According to its own data the most popular device for taking pictures shared on Flickr is the iPhone, beating the Canon EOS, the top digital camera, hands down.

The sharing of camera phone videos sparked an unusual moral panic in Britain in the mid-2000s. Starting in late 2004 and peaking in the summer of 2005, newspapers and TV news programmes carried shocking stories of out-of-control teenagers running up to victims and hitting them while an accomplice filmed the violence on their phone. The practice was called 'happy slapping'. Here is one report by BBC News:

> A 14-year-old has been attacked by three people who videoed the assault on a mobile phone.
>
> The victim and his brother were walking along Dallington Road in Northampton last Wednesday evening when they were approached by three men.
>
> One of the three pushed the boy into a bush before punching him in the side of the head.
>
> The attack was videoed by another man on his mobile phone – a craze known as 'happy slapping'.
>
> Detectives said the three then walked off towards a nearby pub.
>
> One of the offenders has been described as white, between 15 and 20 years old, about 6ft 2in tall and was wearing a white and blue top.
>
> He also appeared to be wearing eyeliner.

In another case the mother of the victim was reported to be demanding that camera phones be banned from schools. The ITV investigative news programme, *Tonight with Trevor MacDonald*, labelled happy slapping an epidemic and sought to place the blame on kids copying violent reality stunt shows such as *Jackass*. In fact it was a moral panic. Camera phones are more widespread now than ever, yet 'happy slapping' no longer troubles the headlines. Partly, according to Graham Barnfield, a lecturer from the University of East London who has ruefully reflected on his unwitting role in the snowballing of the story, the moral panic was caused by a media feeding on itself without taking the time to research further than Google. But partly, the moral panic came from social anxieties, specifically anxieties about the mass extension of making and sharing video images by unsupervised young people. Or, put another way, the furore over 'happy slapping' was an inarticulate and misdirected response to the spread of the intimate personal computer.

The capacity of smartphones to display high quality images, when combined with the ability to determine location, has meant that mobile devices have rapidly displaced the paper map. Just as the mobile phone started as a car-based technology, so mobile cartography first reached public attention in the 2000s as a car-based driving aid. These 'automotive navigation systems', made by companies such as TomTom, Garmin and Navigon, displayed a map of the oncoming road and told the driver where to drive. Digital cartography – such as Google Maps, launched in 2005 – rapidly became standard features on smartphones. They were a spur to further innovation; many applications overlay the basic geographical map with useful information. Google Maps, for example, overlays maps with traffic data – a jam will flash an angry red, a congested road yellow and a clear drive will be a calming green. Cunningly, the data comes from people's phones – it is part of the licence agreement, if you look carefully. By driving around with an Android phone or an iPhone 4 you are feeding location data back to Google which then collates and represents the data back to you as a traffic colour.

The traffic function on Google Maps is a good example of how mobile digital cartography is not merely more convenient than traditional cartography, but represents something qualitatively new. The novelty is not mobility. After all, most paper maps are designed to be mobile. Nor is it simply the overlay of information, although there is no doubt that the ease by which geographical data can be combined digitally is behind the explosion of diverse location-based applications. Rather the revolution comes from the map, once it is held on a smartphone, becoming a portal for information to flow in two directions. Not just from map-maker to map-reader, but vice versa too. Think how the driver consulting Google Maps' traffic report is also feeding back information. Now the map-maker is mapping the map-reader – and changing the map accordingly! The individual gets a useful service. But the company also profits from the new cartography. The aggregate data of movements and locations is a goldmine that map-makers, such as Google and Apple, exploit.

Much of this location-based tracking takes place without the user's attention being drawn to it. Sometimes it can be truly underhand. In the spring of 2011 two programmers, Alasdair Allan and Pete Warden, discovered that the iPhone and the 3G iPad with the operating system iOS4 systematically log the geographical location of the device along with a time stamp in a file called 'consolidated.db'. (Others knew of this file, but it had not been publicised.) As they wrote on their blog, raising the alert: 'Anybody with access to this file knows where you've been.' What is odd is that no one really knew why Apple put it there. For the record, Apple swiftly issued a press release stating that the 'iPhone is not logging your location' – it was just logging the dates and positions of lots of sites *very near* your location, and anyway it was just doing so to make your iPhone work better.

More generally, the collection of location information, made possible either by GPS or by phone mast tracking, has opened up immense sets of data for social and natural scientists to examine. In 2008 the scientific journal *Nature* carried a paper with the title 'Understanding individual human mobility patterns'. In it scientists Marta Gonzalez, C.A. Hidalgo and Albert-Laszlo Barabasi reported on the results of tracking 100,000 anonymised individuals as they carried mobile phones over a six-month period. People generally didn't stray more than ten kilometres from their bases, while a few roved over hundreds. This might not seem to be the most surprising scientific result of all time, but, as of 2012, according to Google Scholar the paper had been cited over 1,000 times. The finding informed work on urban planning, the sociology of friendship and the spread of viruses, both human and computer.

Taking pictures and reading maps are, nevertheless, relatively minor uses of the smartphone. Research on British smartphone users, commissioned by O_2 and released in June 2012, found that on average 25 minutes per day were spent browsing the internet, eighteen minutes checking social networks, fifteen minutes listening to music and fourteen playing games. Making calls was only the fifth most common use – just twelve minutes. Confirmation, perhaps, that we should stop calling these things 'phones'. Taking photographs took up just over three minutes – less even than the nine minutes absorbed by reading a book. In between were activities such as checking and writing emails (eleven minutes) and watching TV and films (nine minutes). In total, this group of smartphone users spent just over two hours a day in constant touch. Most used their phone as their alarm clock. Market research in the United States and Canada reveals broadly similar patterns.

Playing games is an area of culture that has been transformed by the smartphone, and in an extraordinarily short period of time. Reading reports as recent as 2005 and 2006, one is struck by how pessimistic the mobile games industry was. Developers (typically very small new technology companies) had a poor relationship with 'publishers', the mobile companies, in which there wās no agreement about a fair way to channel revenue and divide up profits. And players weren't interested. In the United States in 2006 in an average month, for example, less than 4 per cent of mobile phone users downloaded a game. But the smartphone, and the iPhone model of the App Store in particular, offered solutions to the revenue and quality issues.

The games that have sold well are simple and addictive. Angry Birds involves pinging birds with catapults at the ramshackle defences erected by egg-stealing pigs. It's colourful and equally amusing to nine- and 90-year olds. Fruit Ninja, reputedly a favourite way to 'chillax' of the British prime minister David Cameron, involves sweeping your finger – a ninja's sword – through fruit. It's not complicated. Typically of smartphone culture these games are absorbing – never has constant touch been better illustrated – and only apparently individualistic. One plays Angry Birds alone, in a private bubble, but it has been picked out and purchased because of the social aggregating of consumer choices choreographed by the App Store.

The successful games make their money through

direct sales (59 pence or 99 cents through the App Store, say) and advertising (Android advertising alone is worth $1 million a month to Rovio, maker of Angry Birds). Many also encourage users to link up, for example reporting high scores to social media. As noted above, using social media is among the top uses of smartphones. Facebook, which started in 2004 and has a billion members worldwide, and Twitter, even younger (b. 2006), are currently the leading sites. Yet again we notice the characteristic intimate, personal, individualistic surface appearance which hides a business model based on the gathering and exploitation of aggregate information.

Chapter 28
Cellular war

In March 2003, the United States, in a coalition which included the United Kingdom, invaded Iraq on the pretext of Iraq's failure to satisfy United Nations resolutions regarding the possession of weapons of mass destruction. The war was actively opposed by many countries, in particular France, Germany and Russia. For example, in February, France and Germany (with Belgium) had set opposition to the war above the obligations of NATO membership, by maintaining a veto on plans to defend Turkey if attacked by Iraq. France and Germany insisted that policing Iraq with United Nations weapon inspectors was preferable to war. France, Germany and Russia threatened to veto any new United Nations resolution authorising military action. The Europeans were swiftly demonised by the 'hawks' in George W. Bush's administration.

One of the patterns that we have seen so far in *Constant Touch* is that international political relations powerfully shaped the development of the mobile phone, in particular through the negotiation and operation of cellular phone standards. The squabbles over who would win the contracts to build Iraqi mobile systems in the aftermath of war illustrate again this phenomenon. Mobile phones had been (effectively) banned under Saddam Hussein. In July 2003, the Coalition Provisional Authority invited expressions of interest in three Iraqi mobile licences on offer, and laid down strict rules over what kinds of bid were allowed. One of these rules, which stated that no government could 'directly or indirectly own more than 5 per cent of any single bidding company or single company in consortia', seemed to many commentators designed to exclude companies based in continental Europe, in particular Orange and T-Mobile (in which the French and German governments held significant stakes, a legacy of the companies' origins as spin-offs from national telecommunications authorities). The rule also seemed to exclude the neighbouring Arab companies, including MTC-Vodafone in Kuwait and Batelco (Bahrain) which had already rigged up a working emergency system around Basra and Baghdad respectively. (Nor were these newcomers. Batelco had a venerable history in mobile phones: a tiny Batelco cellular network may well have been operating as early as 1978, which would make it one of the first in the world.)

Even more controversial was the decision over whether to adopt GSM or one of its rivals as the second-generation standard for Iraq. Barely a week into the war, the Republican congressman for San Diego, Darrell Issa, denounced the very suggestion of deploying 'a European-based wireless technology known as GSM ("Groupe Spécial Mobile" – this standard was developed by the French) for this new Iraqi cellphone system'. (He was wrong, of course. As we saw earlier, GSM was not developed by the French alone, although it was undoubtedly a *European* project.) Congressman Issa urged the choice of a rival standard, CDMA, which, as we have seen, was developed in San Diego County by a firm with intimate and long-standing links with the US defence industry, Qualcomm.

So the choice was between GSM, used by all neighbouring countries to Iraq, as well as being the standard on which most of the world's cellphones operated, or CDMA, a product and symbol of American security interests. In August 2003, following an outcry from Arab companies, the 5 per cent rule was relaxed to allow a 10 per cent stake. This still excluded companies such as Batelco. However, GSM was chosen. In the event, the three licences were finally awarded in October 2003, after delay, obfuscation and allegations of corruption, to three Arab consortia: Egypt's Orascom Telecom, Kuwait's National Mobile Telecommunications and MTC (which had strong British links, via Vodafone and the British administration in southern Iraq). As part of the Orascom Telecom deal, the key infrastructure contract went to Motorola, the one company that could boast a happy combination of GSM expertise *and* American ownership.

Once the infrastructure was in place, Iraqi cellular phones were snapped up. By 2012, according to

World Bank figures, 71 subscriptions were in place for every 100 people, which isn't far short of the Middle East average. And the full range of mobile culture and uses developed. Without mobile phones the infamous 2004 images of abused prisoners at Abu Ghraib prison would probably not have been taken, and certainly not as widely shared. Likewise, mobile phone shops in the Shia areas of Baghdad were selling camera phone footage of Saddam Hussein on the gallows within a day of his hanging on 30 December 2006.

In Afghanistan, whose conflict with the West has dragged on just as long as in Iraq, there are other mobile stories to tell. It's a poorer country and the market was less tempting to cellphone companies. Nevertheless, mobile subscriptions by 2012 stood at about four in ten of the population, and four companies combined to offer mobile phone services across three quarters of the country. The fact that cellphones could be tracked or eavesdropped meant that these companies became the target of the Taliban. In February 2008, a Taliban statement demanded that the companies turned off the cellphone infrastructure from 5.00pm to 7.00am. 'If they do not heed it', ran the statement, 'the Taliban will target their offices, suboffices and tower stations.' The companies agreed. Interestingly the Taliban did not call for a total suspension – presumably the phones are just too useful. But some analysts saw a message in the Taliban's partial blackout as well as a compromise.

'Tactics like the cellphone offensive have allowed the Taliban to project their presence in far more insidious and sophisticated ways, using instruments of modernity they once shunned,' notes Alissa J. Rubin in the *New York Times*. 'The shutoff sends a daily reminder to hundreds of thousands, if not millions, of Afghans that the Taliban still hold substantial sway over their future.'

Afghan workers inspect the burned remains of mobile phone equipment after an attack by Taliban militants, Kandahar, March 2008. It was the second such assault in two days, after the Taliban warned phone companies to shut down their towers or face attacks. (Press Association)

Chapter 29

The revolution will not be mobilised

As mobile phones have become widespread through-out much of the globe, so it is no surprise that they are used as a tool of communication for all purposes, including the organisation of political protest. We have already seen the case of the ousting of Joseph Estrada in the Philippines in 2000–01. But great care has to be taken in the analysis of such episodes, for two reasons. First, commentators in the West have been over-eager to attribute unwarranted, specific power to new tech-nologies as tools of political protest. The novelty can distract attention from the continuing and probably more important roles of older methods of organisa-tion. Second, unpleasant regimes have not been slow to learn lessons. Indeed, there are plenty of reasons to think that new information technologies offer new ways to repress protest and to bolster authoritarianism.

Between 2003 and 2011 the world witnessed a series of protests, many of which subsided but some of which succeeded in ending long-standing repressive regimes. These episodes include the so-called 'colour revolutions' (such as the 'Rose Revolution' of Georgia in 2003 and the 'Orange Revolution' in Ukraine in 2005) and the 'Arab Spring'. In some of these protests the use of mobile phones and social media caught the eye of commentators in the West, who proposed alternative monikers. For example, the Moldovan unrest of 2009, the 'Green Revolution' protests over the disputed elections in Iran between 2009 and 2010, and the revolutions in Tunisia and Egypt between 2010 and 2011 were all called 'Twitter revolutions'. (The Iranian protest, the end of Hosni Mubarak's regime in Egypt and that of Ben Ali's regime in Tunisia were also called 'Facebook revolutions'.)

There is no doubt that the use of mobile phones and social media were part of the story. In Iran, camera phone images circulated on YouTube of the death of a 26-year-old woman called Neda Agha-Soltan contributed immensely to the opposition cause. In Tunisia, likewise, video footage of the small-scale protests early in the revolution was recorded and shared using mobile phones. But even then it depended on older media technologies – specifically a traditional television news network, Al Jazeera – to pick up on and spread these images further to fully ignite the unrest. Yet still it was Twitter, Facebook, YouTube and the mobile phone that were picked out in the West as most significant. Partly this highlighting of just one aspect of the episodes can be explained by a deep-seated set of values that presents new information technologies as inherently revolutionary. Partly it comes from a utopian belief, dating from the Cold War, that free information alone can undermine authoritarian regimes. Partly it was simply a result of reliance by journalists and commentators in the West on what was visible and accessible to them. Faced with fast-moving and dangerous situations in far-away countries, it was easier to monitor Twitter feeds (especially English-language Twitter feeds) than pursue the stories on the ground.

Indeed the more closed a society is to the West, the more likely it is that small, visible, accessible trickles of information are seized upon and their significance overplayed. For example, take North Korea, undoubtedly the most closed country on the planet. In 2010, the *New York Times* carried the headline 'North Koreans use cellphones to bare secrets'. On one level the story was about the rise of websites collating anonymised news of everyday life in the communist paradise, drawn from North Koreans who bravely sent them information. In fact the article, by Choe Sang-Hun in Seoul, reveals just how effective the regime was in stopping this flow. Very few North Koreans have mobile phones. The only area where messages can be sent is in the far north, where the Chinese cellular networks just about reach across the border. Genuine whistleblowers put themselves in immense danger (and if caught were shot). The North Korean regime, well aware of what the websites were doing – and the fact that three of the five websites involved were funded by the American National Endowment for Democracy surely helped it interpret their activities ideologically – has responded in a depressingly sophisticated manner, feeding disinformation and monitoring calls.

The scholar Evgeny Morozov is the most trenchant, and insightful, critic of naïve and over-optimistic accounts of the relationships between new information technologies and authoritarian regimes. In *The Net Delusion* (2011) he points out, for example, the lack of hard evidence that Twitter was the critical organising tool of the Iranian protests, but also how this absence of evidence did not prevent the claim being repeated across the blogosphere. (Indeed he cites Moeed Ahmad, director of new media for Al Jazeera, who conducted a fact-checking exercise and could 'confirm only 60 active Twitter accounts in Tehran, a number that fell to six once the Iranian authorities cracked down on online communications'.) Morozov rightly says that 'the mobile phone is another activist tool that has not been subjected to thorough ... analysis', one which has plenty of 'shortcomings and vulnerabilities'. Specifically, he notes that mobile phones play into the hands of regimes. First, they can turn off cellular networks, either across a nation or just in sensitive areas. For example, the Belarus government in 2006 and the Moldovan authorities in 2009 simply ordered the shutting down of the mobile network in the central squares where protestors were gathering. Second, Morozov finds plenty of evidence that authoritarian regimes are becoming masters of sophisticated keyword searching, turning rebellious SMS messages into a mine of useful information for surveillance and harassment. In China, mobile operators routinely survey and block messages with banned words, while China Mobile's chief executive officer freely admitted in 2008 that 'his company provides data on its users to the government whenever the government demands it'. Finally, Morozov argues that location-tracking makes this capacity to use technologies against protestors and dissidents even more powerful, further strengthening authoritarian states.

Chapter 30
Oases of quiet

The days when mobile phones were a symbol of privilege are long gone. In many countries the number of subscriptions equals or even exceeds the population. People are talking more, at least on the cellphone. The soundscape has also changed. In the 1990s the distinctive 'Nokia Tune' – de, de, duh-dah, de, de, duh-dah, de, de, duh-dah-dah, in gently falling tones; in fact a few bars of the Spanish guitar composer Francisco Tárrega's *Gran Vals* of 1902 – rang out. (In the future, directors will be able to evoke the last years of the second millennium just by playing this phrase). With first customisable ringtones and then the infinite sonic potential of smartphones, the sonic soundscape of the 2000s is perhaps less distinctive. While teenagers on the top floor of a bus might indulge in 'sod-casting', the playing of music at loud volume through the tinny speakers of their phones, most of the time on a smartphone, as we have seen, is spent silently searching the internet, updating social media and playing games. And this is done almost everywhere there are people.

Nevertheless it is now far easier to identify the places that don't have cellphones than to attempt to describe all the places that do. That said, if you look at a map of global GSM coverage – a good indicator, since it is the most popular and cheapest standard – then there are plenty of white areas where there is no signal. The upper Amazon, the Sahara, the drier lands of west China and central Asia and the Australian outback are all blank. And of course the frozen Arctic and Antarctic are not covered with cellphone masts. Cellphone coverage is spotty on a finer grain too; it peters out as population and wealth dwindle. Where there are people, it seems, there are phones. However, the exceptions are fascinating, because they reveal the social and technical rules which constrain the mobile phone.

There are some rooms where you should not use a cellphone. In a gym, a camera phone is an invasion of privacy. Elsewhere it's a problem of disturbance. In theatres and concert halls, for example, no one wants to have a performance interrupted by a ringtone, a sound always followed by the noise of scrabbling as the embarrassed punter tries to locate the phone in the dark and fumbles to turn the thing off. Yet, of course, with bigger audiences the chance that one person has forgotten the switch off increases towards inevitability. It is a gruesome breach of social etiquette, at least in the politer halls of entertainment. In November 2005 the actor Richard Griffiths was towards the end of a Saturday matinée performance of *Heroes*, a play about war veterans in a nursing home, when an audience member's phone went off repeatedly. The first time, the actor just gave the offending woman a withering stare.

The third time, Griffiths, whom younger readers might know as Mr Dursley in the Harry Potter films, angrily told her to turn it off or leave. The audience gave a deafening round of applause – a sure sign that a well-known social rule had been re-established. 'It's a phone-free zone. We don't want them ringing and we certainly don't want them ringing and people ignoring them pretending that it's not theirs,' said Kevin Spacey, star of *The Usual Suspects* turned London theatre manager, of an earlier incident, adding: 'My feeling is if people don't know how to behave they shouldn't come.'

In Britain, phones and BlackBerrys are banned from the meetings of David Cameron's Cabinet, partly because they are a distraction but also to reinforce a constitutional point that ministers share responsibility for decisions and therefore need to pay attention. Phones in Parliament are allowed, so long as they are set on silent. At the top, phones are banned because they interfere with the operation of power and authority. At the bottom, phones are banned to make people powerless. In the United States justice system, for example, prisoners in all state and federal prisons are forbidden to possess a cellphone – pun intended. But as Kim Severson and Robbie Brown revealed in the *New York Times*, the reality is quite different. 'Almost everybody has a phone,' said 'Mike', an inmate at Smith State Prison in Georgia, to the reporters. 'Almost every phone is a smartphone. Almost everybody with a smartphone has a Facebook.' It is not gentle Facebook poking that worries authorities. Rather, since the smartphone is a personal computer connected to the internet, it can be used to access all kinds of 'files', and not of the file-in-a-cake variety. Rather the data could be maps, instructions to criminal compadres or means of organising a prison protest. Nor can cellphone signals be jammed, since this would interfere with the constitutional right to communicate held by those with legal phones – lawyers, guards – who visit prison.

The banning of mobile phones is, significantly, most earnestly wished for in institutions that depend on a well-defined and strong hierarchy of authority. Prisons, hospitals and courtrooms are exemplary cases. Anxiety over phones in schools also partly springs from this source. Mobile phone bans are a matter of individual school policy in Britain, but in Osaka in Japan, New York City and the state of Bavaria in Germany, for example, there is a region-wide prohibition. Time spent texting is time not spent paying attention to the teacher. Ringtones are as disruptive in the classroom as they are in a hushed theatre. However, there is a mismatch between official policies and actual practice. The Pew Internet and American Life Project surveyed the use of cellphones by teens in the United States in 2009, and paid particular attention to schools. The majority of pupils attended schools that permitted cellphones to be brought to school but forbade them in class. A quarter banned cellphones altogether, while a mere one in ten were permissive. However even in the no-phone schools, over six out of ten teens smuggled their cellphone in every day. Phones were hidden 'behind stacks of books, under desktops, inside of bags', while one pupil was even more inventive: 'I've got [a second phone] ... if you get caught using your phone you can pull out a fake phone, turn it on and give it to them.'

Pupils want phones because they are an important tool for maintaining their social lives. Parents sometimes side with their children rather than with the school in this area. New York is an interesting special case because the city has a rule, introduced in the 1980s at the time of pagers, that bans all electronic communications devices from school premises. 'Schools are for learning,' said the city's mayor Michael Bloomberg in 2006, backing the ban, 'and these devices are diversions from learning.' But parents disagreed. 'As a single parent of three children, at three different levels, in three buildings, I have no choice but to use a cellphone to coordinate logistically,' said Carmen Cola. Another parent added: 'If [my child] couldn't take his cellphone, I wouldn't want him travelling to school in Manhattan ... I'm going to figure a way to hide their cellphones so that they'll have it.' One entrepreneurial response to the prohibition was devised by convenience stores near school gates – they installed safes so that pupils

A sign banning mobile phone use is installed outside a school in Austin, Texas, 2009. (Press Association)

could leave their phones in the morning and pick them up on the way home.

Another set of buildings that restrict the presence of mobile phones is corporate headquarters and laboratories. The worry here is over commercial secrecy. Apple has always fiercely protected its creative back spaces. You would never be allowed to take an iPhone into the place where iPhone prototypes are designed, played with and tested. Other companies – such as Intel and Samsung – prohibit phones, especially camera phones, from their premises. Commercial secrecy shades into national security too – phones are banned from sensitive sites such as the Lawrence Livermore nuclear weapon design centre in the United States.

Two technologies of mobility, cars and mobile phones, have developed together, but the intimate relationship might be unravelling with another set of restrictions and prohibitions. One of the first ways in which radio telephones were used was in police patrol cars in American cities. The commercial mobile radio services of the 1950s and 1960s were marketed at the chauffeur-driven classes or at businesses where mobility was part of the job. Even Lars Magnus Ericsson, back in 1910, first imagined a mobile telephone while driving in the Swedish countryside. Furthermore, both technologies have featured a Janus-faced aspect: on the one hand they have been used to powerfully express and reinforce the freedoms of the individual, while on the other the centralised databases that accompany them – the driving and vehicle licence databases and the information kept about cellphone subscribers' movements, actions and contacts – have been used by the forces of law and order. Technologies of mobility created new opportunities for crime, but also the tools by which crime was fought. In the United Kingdom, a new twist in the relationship occurred on 1 December 2003 when it became a specific offence to use a hand-held phone when driving. Similar bans were introduced in countries from Australia and Austria to Turkmenistan and Zimbabwe. In the United States, like the mobile service itself, the regulations were patchy – total bans in New York and California, partial bans in Florida and

Illinois, but none in Alabama and Idaho. While most 'hands-free' phones are still legal, the bans do seem to be the beginning of the end of the affair.

Anyone who has flown in an aircraft will be familiar with the following announcement prior to take-off: 'At this time, we request that all mobile phones, pagers, radios and remote-controlled toys be turned off for the full duration of the flight, as these items might interfere with the navigational and communication equipment on this aircraft.' The request is often frustrating and sometimes – as in the case of the car – ignored, both signs of how we now expect to be in constant touch. (In 2006, the Federal Aviation Administration discreetly surveyed 38 flights for cellphone calls, and found 'considerable onboard radio frequency activity'.) The problem is that aircraft are complex assemblages of electronic and other systems, which can be affected in unpredictable ways by signals from mobile phones. When switched on, the phone attempts to connect to a network and, not finding one, boosts its signal to maximum power. In 2008 the RTCA (originally the Radio Technical Commission for Aeronautics, but which acts as an independent body bringing together industry, government and academic parties), found that unauthorised electronic emissions 'could exceed interference thresholds for critical aircraft systems', notably two systems: one for determining position, one for recording the glide slope, the angle crucial to landing.

However, there is money to be made from passengers desperate to stay in touch with the ground. The European plane maker Airbus announced research in 2006 that claimed that mobile phones could be used safely in flight. There remains the problem of distance – 35,000 feet is a long way from the cellular networks. (Indeed, reception is spotty just at the top of city skyscrapers.) In the Airbus trials the A320 airplane carried a 'picocell' (a tiny cell), which communicated to Globalstar satellites, and thence to the ground. A similar, contemporary trial was carried out by American Airlines in collaboration with Qualcomm. In 2009, the budget operator Ryanair – never one to miss an opportunity to find new things to bill passengers for – introduced a chargeable service allowing calls to be made over 10,000 feet. Other airlines – Qantas, Malaysia Airlines, Emirates and Royal Jordanian – did the same. The ban, however, remained in the United States, where cabin crew insisted that cellphones would aggravate the already high tensions between passengers and would interfere with the need to give clear direction in emergencies.

Phones on underground trains tell a similar story. The problem has been both technical – how can a phone be made to connect to the cellular network? – and social – will other passengers object? The technical problem is not merely distance from the ground, since base stations can easily be installed in the tunnels.

Rather the problems are ones of speed and capacity. 2G and 3G networks do not work well with fast-moving phones, and an underground train carriage can cram together 100 passengers, each with a phone. Nevertheless, these technical difficulties can be overcome – as they were on the Paris Métro (Europe's first to work with cellphones) and even the Tyne and Wear Metro in 2004. In London, progress is slow. One reason for delay is concern that mobiles might be used to trigger terrorist bombs. 'Using mobiles in deep line sections' was not necessary, said Simon Hughes, a London mayoral candidate, in 2005, adding: 'Texting is a luxury, security is not.' In 2012 there was still no signal on the Tube, with no agreement on who should pay for the installation if it happened: taxpayers and passengers or mobile operators.

For some the continued silence is welcome. But the case of the BART unrest shows that passengers, once connected, would soon expect that to continue. BART is the Bay Area Rapid Transit system, the tram that runs in and out of San Francisco. In July 2011, protests were sparked by the killing of a homeless man who was shot by a BART police officer at Civic Center station. Twitter (using the hashtag #MuBARTek, in reference to the ex-leader of Egypt) was one tool used to coordinate action. As the situation escalated, the BART authorities decided to turn off the cellular network to thwart the organisation of the protestors. This in turn caused general outrage from commuters, and even provoked cyber attacks, supposedly by the Anonymous hackers' collective. One side cried 'freedom of speech' while the other said this was trumped by public safety concerns.

The BART episode is a curious one. The denial of a cellphone service – the temporary blackout – is a tool of authorities, but a blunt one. Similar examples can be found across the world. Just after the London 7/7 bombings in 2005, the cellphone service in four of Manhattan's tunnels – Holland, Lincoln, Midtown and Battery – was switched off for several days. In the same year in Thailand, where the government was facing an insurgency in the Muslim-majority south of the country, unregistered prepaid phones were disconnected. Again the concern was terrorism, specifically the use of phones as triggers. India did the same to millions in 2009, as we have seen. In the Madrid train bombing of 2004, mobile phones were used as timers to trigger the bomb – essentially using the phone as a clock rather than as a communication device.

So we have seen examples of many different kinds of 'oases of quiet', from single rooms to whole buildings, tunnels and railway tracks. But what about whole countries? Until recently, only government officials or people working for foreign firms could purchase a mobile phone in Cuba. In April 2008, Raul Castro, the new president of the country and younger brother of Fidel, lifted the restrictions. Crowds gathered at shop windows on the first day. Within ten days, 7,400 new mobile subscriptions had been made through the state telecoms firm, Etecsa. In reaction, scenting an opening, the United States allowed relatives to send mobile phones by post to the communist island. Usage in Cuba remains low, however: a consequence of poverty now rather than prohibition.

And there will always be people who do not want a mobile phone. They will preserve their own oasis of quiet, and quite rightly so. One personal response springs to mind. It is a vox pop comment to a BBC News feature on the iPhone: the words of Tracy Churchill, from Ely in the Cambridgeshire fenland:

> iThis and iThat ... What is happening to the world? Have we lost the art of entertaining ourselves without digital gadgets? I am proud to say I do not own a mobile, let alone one with camera, internet access and a million other things that cost people money, but don't use. If someone wants to speak to me they can use my landline, or write me a letter ... old-fashioned maybe, but at least I'm not permanently attached to an inanimate object isolating me from the wonderful natural world around us.

Chapter 31
Perpetuum mobile?

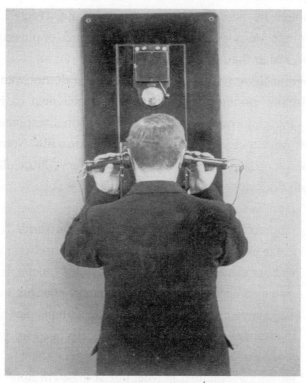

(BT Archives)

By 2002 there were at least a billion cellular subscribers in the world. By 2012 there were six, including a billion smartphones alone. The mobile cellular phone has meant different things to different people: a way of rebuilding economies in eastern Europe, an instrument of unification in western Europe, a fashion statement in Finland or Japan, a mundane means of communication in the United States, or an agent of political change in the Philippines. Different nations made different mobiles.

But, through all the contrasting national pictures of cellular use, a common pattern can be glimpsed: there has been a correlation, a sympathetic alignment, between the mobile phone and the horizontal social networks that have grown in the last few decades in comparison with older, more hierarchical, more centralised models of organisation. There is, I feel, a profound sense in which the mobile represents, activates, and is activated by these networks in the way that, say, the mainframe computer of the 1950s gave form to the centralised, hierarchical, bureaucratic organisation. What changed between then and now was a social revolution – of which technological change was part and parcel. The critical attitude towards centralised authority that emerged in the 1960s can be seen in examples as diverse as the CB radio fad of the early 1970s, with its alternative jargon and myths of living outside traditional society, or the use of mobile phones to organise illegal rave parties around London's M25 motorway in the late 1980s and early 1990s. Social revolution produced the youth movement that in turn demanded a material culture – including cellphones – that would meet young people's social needs of *distinct* fashions, and *independent* means of communication between friends.

We must be careful not to suggest that technologies of mobility inevitably oppose centralised power. Just over 2,000 years ago, city-states around Italy and the Mediterranean began to feel the force of a new power: imperial Rome. These city-states came to accept – and see – themselves as 'Roman' partly because 'Roman' roads gave the social elite an effective means of travel. Anywhere visible from the road became somewhere from which this association of mobility, the technologies enabling mobility, and centralised Roman power could be reinforced. The great triumphal arches, adverts for 'Rome' through which a traveller had to pass, provide one example. As political power became concentrated in the hands first of an oligarchy and senate, and then of individual emperors, so the road system became ever more important. Technological systems of mobility helped create the Roman world, and mobility reinforced central, hierarchical, imperial power.

But try a historical thought experiment: what would Rome have been like if the technologies of mobility had been truly demotic? If, instead of being built to create and sustain the mobility of a social elite, the roads and road culture of Rome had been for the people? Well, I would suggest as a parallel that the mobile phone as it became in the last decade of the 20th century, and the car by mid-century, are just such demotic – and interconnected – technologies of mobility. GSM, the 'most complicated system built by man since the Tower of Babel', was meant to make 'Europeans' as the roads made 'Romans', but it was not dedicated to the maintenance of a privileged elite. (However, in other ways the comparison stands. The power of several mobile-based multinationals is becoming imperial in stature, and operator portals, which advertise the riches of such companies and through which we must pass, bear more than a passing resemblance to triumphal arches.)

There are fierce tensions between demotic technologies of mobility and centralised power. I've given many throughout this book, but the thought of Rome prompts one more example. In March 2001, the Italian Bishops' Conference, the governing organisation of the Roman Catholic Church in that country, circulated to all parish priests a strongly worded warning not to allow mobile aerial masts on church buildings. The mobile operators were in desperate need of sites for new masts, preferably high up and in the centres of populations, and were willing to pay handsomely. The parish priest, often some distance from the riches of the Vatican, had an expensive church to maintain and falling congregations. While some very worldly concerns fell against the obvious deal – such as legal violations that might endanger churches' tax-exempt status – a greater danger arose: the centrality of Christian symbols would be blurred if the spires also advertised mobile masts. The masts, wrote the bishops, were 'alien to the sanctity' of churches. And in a world where the Catholic church, perhaps the prime model of authority and hierarchy, perceived numerous threats, the blurring of symbols of mobility and static power was unconscionable. (Broadcasting, always at ease with hierarchy, was another matter: Vatican Radio was permitted its masts.) In contrast, the more compromising, more pragmatic or less symbolically minded Church of England signed a deal with Quintel S4, a spin-off of QinetiQ (the commercial arm of Britain's defence research agency) in June 2002 to allow mobile masts on 16,000 churches.

But if compromises can be made with hierarchies, the demotic mobile has been found to fit within horizontal social networks with greater ease. This was not discovered by the mobile phone manufacturing or operating companies (although companies such as Nokia can boast a flattened management structure, and espouse a pro-innovation, anti-deferential corporate philosophy). Instead, the demotic mobile was the discovery, and reflection, of the users. No one within the industry, for example, expected the extent of the success of Short Message Service (SMS). Indeed, even now it is remarkable that a service that works out on average at roughly a penny or cent *per character* was successful at all. But the power of text, like many other aspects of the mobile, was found by the people who used it, not the people who planned it.

The mobile phone in the early 21st century is in a moment of transition. The third generation has been launched and a fourth is being rolled out; there are rival models of wireless communication, some centralised (like the satellite system Globalstar), some even more potentially demotic (wireless LAN, Bluetooth and a host of other means of passing data from device to device free of charge). The mobile – as a phone – is in danger on three fronts: technological change might add to it so many new features, benefiting from greater data-handling capacities, that it might barely act like a phone. Its own flexibility would destroy it, or transform it into something else. The mobile phone would have been a mere passing stage to another technology. Perhaps some rival means of mobile communication and data handling across ad hoc networks will prove more economical, more popular or a better fit for social or political imperatives. The mobile cellular standards will lie abandoned. The immense sums spent on outmoded or unwanted 3G and 4G licences will collapse some of the biggest corporations, with significant effects on the rest of the economy. There is real nervousness that the new mobile might be rejected by a more powerful force, the users. Certainly patterns of use are changing as the smartphone displaces the older cellphone. The number of 'calls', for example, actually

fell for the first time in Britain in 2011, as voice communication becomes less central. 'Teenagers and young
adults are leading these changes in communication
habits, increasingly socialising with friends and family
online and through text messages despite saying they
prefer to talk face to face,' concluded Ofcom, Britain's
regulator watchdog.

But there is good reason to suggest that users will
continue to love the mobile phone. Back in 1906, the
inventor of electronic circuitry, Lee de Forest, made
the first radio transmission to an automobile. A press
release issued to advertise the achievement expressed
de Forest's hope that in the near future 'it will be possible for businessmen, even while automobiling, to be
kept in constant touch'. In the centralised, hierarchical
world of Ford, such a dream made little sense. While
it was technically possible earlier, the cellular mobile
phone only took off after the 1960s. By then there had
been a sea change in both politics and technology, one
affecting the other; after which, to put it crudely, networks of people have prospered and hierarchical styles
have suffered. When I smashed up my mobile phone
I wanted to find out what was in it and in what sort of
world it made sense to assemble it. Apart, the debris
reflected a fragmented, flexible, atomised world. Put
together, the mobile provides a *network*, giving society
back a cohesion of sorts. We've seen that the phone can
be assembled in many ways. But only after the great

transformation of social attitudes that took place in the 1960s could a world *wish* to be in constant touch. We live *this* side of that transformation, which is why the mobile will be, if not perpetually in motion, at least moving for some time yet.

And great claims are made about such a world. With the first and second generation of mobile phones, constant touch meant constant communicative contact. With smartphones, constant touch means this and more: stroking screens, moving data and keeping in contact via social media. 'Mobile communication has arguably had a bigger impact on humankind in a shorter period of time than any other invention in human history,' suggested the authors of *Maximising Mobile*, a World Bank report on the opportunities offered for development. They quote, approvingly, Jeffrey Sachs, who directed the United Nations Millennium Project: 'Mobile phones and wireless internet end isolation, and will therefore prove to be the most transformative technology of economic development of our time.' But there are two very different conclusions drawn about whether this world is converging or not. Steve Jobs, as reported by his biographer Walter Isaacson, describes a moment when, in Istanbul and listening to a talk on the history of Turkey, he thought it was:

The professor explained how the coffee was made very different from anywhere else, and I realized,

'So fucking what?' Which kids even in Turkey give a shit about Turkish coffee? All day I had looked at young people in Istanbul. They were all drinking what every other kid in the world drinks, and they were wearing clothes that look like they were bought at the Gap, and they are all using cellphones. They were like kids everywhere else. It hit me that, for young people, this whole world is the same now. When we're making products, there is no such thing as a Turkish phone, or a music player that young people in Turkey would want that's different from one young people elsewhere would want. We're just one world now.

I respectfully disagree. Turkish youth will find their own ways to use Turkish phones. The stories told in this book show how diverse mobile technologies and cultures have been. The global trade routes are extraordinary – recall the case of American phones being recycled by Pakistani wholesalers based in Kowloon, and then being purchased by Nigerian middlemen for their home market. But also recall that when the phones reached rural Nigeria they were used in ways quite distinct from the original American patterns. We may all be in constant touch but that does not mean the world is flat.

Bibliography

Alasdair Allan and Pete Warden, 'Got an iPhone or 3G
iPad? Apple is recording your moves. A hidden file in
iOS 4 is regularly recording the position of devices',
O'Reilly, 20 April 2011 http://radar.oreilly.com/2011/04/
apple-location-tracking.html

Tatum Anderson, 'Mobile phone lifeline for world's poor',
BBC News, 19 February 2007 http://news.bbc.co.uk/1/
hi/business/6339671.stm

Tatum Anderson, 'Mobile phones reach Uganda's villages',
BBC News, 15 November 2007 http://news.bbc.co.uk/1/
hi/business/7071636.stm

Tatum Anderson, 'India seeks mobiles for the masses', BBC
News, 2 December 2007 http://news.bbc.co.uk/1/hi/
business/6339519.stm

Anon, 'From iPhone to iGroan', BBC News, 12 January 2007
http://news.bbc.co.uk/1/hi/magazine/6252991.stm

Artur Attman et al, *L.M. Ericsson 100 Years*, three volumes,
(Örebro, 1977). Three-volume company history of
Ericsson

Ken Banks, 'Mobiles offer lifelines in Africa', BBC News,
15 September 2009 http://news.bbc.co.uk/1/hi/
technology/8256818.stm

David Barboza, 'In China, knockoff cellphones are a hit',
New York Times, 28 April 2009

Anne Barnard, 'Growing presence in the courtroom:
cellphone data as witness', *New York Times*, 6 July 2009

Graham Barnfield, 'How I unwittingly helped to start
the Happy Slaps panic. A modern media tale: my
15 minutes of fame commenting on those 15-second

videos', 20 May 2005 http://www.spiked-online.com/
Articles/0000000CAB59.htm

Nick Bilton, 'Disruptions. At its trial, Apple spills some
secrets', *New York Times*, 5 August 2012

Ryan Block, 'Live from Macworld 2007: Steve Jobs keynote'.
http://www.engadget.com/2007/01/09/live-from-
macworld-2007-steve-jobs-keynote/

Barry Brown, Nicola Green and Richard Harper (eds.),
*Wireless World: Social and Interactional Aspects of the
Mobile Age* (London: Springer, 2002)

Oliver Burkeman, 'How Google and Apple's digital
mapping is mapping us', *Guardian*, 28 August 2012

Business Week: i-mode quotations taken from http://www.
businessweek.com/adsections/sun/heroes/content.
html

Manuel Castells et al, *Mobile Communication and Society:
A Global Perspective* (Cambridge, MA: MIT Press, 2007)

Rory Cellan-Jones, 'Phone hacking: are you safe?', BBC
News, 12 July 2011 http://www.bbc.co.uk/news/
technology-14118995

Cellular News: An excellent resource, at www.
cellular-news.com

Choe Sang-Hun, 'North Koreans use cellphones to bare
secrets', *New York Times*, 28 March 2010

CNN: For the Congo volcano and mobile phones.
For example http://www.cnn.com/2002/WORLD/
africa/01/20/congo.mobiles/

Council of the European Communities, '87/371/EEC:
Council Recommendation of 25 June 1987 on the
coordinated introduction of public pan-European
cellular digital land-based mobile communications in

the Community'. This and other European documents
can be found via europa.eu

Jon Cronin, 'A mobile vision for Africa', BBC News, 5 July
2004 http://news.bbc.co.uk/1/hi/business/3854495.stm

Jon Cronin, 'Africa's mobile entrepreneurs', BBC
News, 24 January 2005 http://news.bbc.co.uk/1/hi/
business/4145435.stm

Mirjam de Bruijn, Francis Nyamnjoh and Inge Brinkman
(eds.), *Mobile Phones: The New Talking Drums of
Everyday Africa* (Leiden: African Studies Centre, 2009). A
rich collection of essays on mobile Africa. Essays discuss
aspects of mobile phone use and culture in Cameroon,
Tanzania, Zanzibar, Sudan and Ghana

Charles Duhigg and David Barboza, 'In China, human
costs are built into an iPad', *New York Times*, 25 January
2012

Charles Duhigg and Keith Bradsher, 'How the US lost out
on iPhone work', *New York Times*, 21 January 2012

Zusha Elinson, 'After cellphone action, BART faces
escalating protests', *New York Times*, 20 August 2011

Kristi Essick, 'Guns, Money and Cellphones', *The Industry
Standard*, 11 June 2001. An informative article on coltan.
Also useful are the United States Geological Survey
reports available via the USGS website, www.usgs.gov

Jonathan Fildes, 'Mobile phones expose human habits',
BBC News, 4 June 2008 http://news.bbc.co.uk/1/hi/
7433128.stm

Claude S. Fischer, *America Calling: A Social History of the
Telephone to 1940* (Berkeley: University of California
Press, 1992). This is the main reference for the history of
the old telephone

H.N. Gant, *Mobile Radio Telephones: An Introduction to*

their Use and Operation (London: Chapman & Hall, 1959)

Garry A. Garrard, *Cellular Communications: Worldwide Market Development* (Boston: Artech House, 1998). The most useful guide to early mobiles. Rich source of information, especially on the technical side, and for international comparisons

Juliette Garside, 'Apple's efforts fail to end gruelling conditions at Foxconn factories. Abuses continue at electronics assembly firm with staff working up to 80 hours' overtime a month, says Hong Kong rights group', *Guardian*, Wednesday 30 May 2012 http://www.guardian.co.uk/technology/2012/may/30/foxconn-abuses-despite-apple-reforms

Gerard Goggin, *Cellphone Culture: Mobile Technology in Everyday Life* (London: Routledge, 2006). Excellent guide to the cultural significance of mobile phones. Topics covered by Goggin include mobile phones as badges of identity, text messaging, cameraphones and mobile television. Particularly noteworthy is a very good chapter on cellphones and disability issues

Fiona Graham, 'M-Pesa: Kenya's mobile wallet revolution', BBC News, 22 November 2010 http://www.bbc.co.uk/news/business-11793290

Louise Greenwood, 'Africa's mobile banking revolution', BBC News, 12 August 2009 http://news.bbc.co.uk/1/hi/business/8194241.stm

Victoria Harrington and Pat Mayhew, *Mobile Phone Theft*, Home Office Research Study 235, Home Office Research, Development and Statistics Directorate, January 2001

Heather A. Horst and Daniel Miller, *The Cellphone: An Anthropology of Communication* (Oxford: Berg, 2006)

iFixit, 'iPhone 4 teardown' http://www.ifixit.com/Teardown/iPhone+4+Teardown/3130/1. I couldn't afford to smash my own iPhone 4, but here is a detailed description of what you see if you strip one down to its components

Walter Isaacson, *Steve Jobs* (London: Little, Brown, 2011). An excellent biography that has as good access to its difficult subject as could be expected

Mizuko Ito, Daisuke Okabe and Misa Matsuda (eds.), *Personal, Portable, Pedestrian: Mobile Phones in Japanese Life* (Cambridge, MA: MIT Press, 2005). An edited collection which takes a snapshot of Japanese mobile culture

A. Jagoda and M. de Villepis, *Mobile Communications* (Chichester: John Wiley, 1993). Published in France in 1991. A good source for the European factors behind GSM

James E. Katz and Mark A. Aakhus (eds.), *Perpetual Contact: Mobile Communications, Private Talk, Public Performance* (Cambridge: Cambridge University Press, 2002)

Timo Kopomaa, *The City in Your Pocket. Birth of the Mobile Information Society* (Helsinki: Gaudeamus, 2000). Finnish social research, confident that the mobile is reinvigorating public space

Ray Lawrence, *The Roads of Roman Italy: Mobility and Cultural Change* (London: Routledge, 1999). Roads and an 'alteration in the mentalité of space-time' to produce Romans

Amanda Lenhart, Rich Ling, Scott Campbell, Kristen Purcell, *Teens and Mobile Phones*, Pew Internet and American Life report, April 2010 http://www.pewinternet.org/Reports/2010/Teens-and-Mobile-Phones.aspx

The Right Honourable Lord Justice Leveson, *An Inquiry into the Culture, Practices and Ethics of the Press*, Four Volumes, HC 780-I (London: The Stationery Office, 2012). This is the Leveson report. The inquiry was triggered by revelations about mobile phone hacking, but covered broad questions about the ethics of the press.

Mail on Sunday, 'The stark reality of iPod's Chinese factories', 18 August 2006 http://www.dailymail.co.uk/news/article-401234/The-stark-reality-iPods-Chinese-factories.html

Mobile Communications International: A good source of mobile news and international statistics

Jon Mooallem, 'The afterlife of cellphones', *New York Times*, 13 January 2008. Very interesting article on the recycling of phones

Olga Morawczynski and Mark Pickens, 'Poor people using mobile financial services: observations on customer usage and impact from M-PESA', CGAP, August 2009

Evgeny Morozov, *The Net Delusion* (London: Allen Lane, 2011). Thorough-going and refreshing scepticism

Robert C. Morris, *Between the Lines: A Personal History of the British Public Telephone and Telecommunications Service, 1870–1990* (Just Write Publishing Ltd, 1994)

O_2, 'Making calls has become fifth most frequent use for a Smartphone for newly-networked generation of users', http://news.o2.co.uk/?press-release=making-calls-has-become-fifth-most-frequent-use-for-a-smartphone-for-newly-networked-generation-of-users

Ofcom, *The Communications Market Report: United Kingdom*, 2012 http://stakeholders.ofcom.org.uk/market-data-research/market-data/

communications-market-reports/cmr12/uk/. This
report found 'for the first time ever, a fall in the volume
of mobile calls (by just over 1%) in 2011'. Also contains
good data on other mobile consumer trends

Public Record Office: HO/255 series has many interesting
files relating to early mobile radio in the UK

Richard Robison and David S.G. Goodman, *The New Rich
in Asia: Mobile Phones, McDonalds and Middle-Class
Revolution* (London: Routledge, 1996)

Alissa J. Rubin, 'Taliban using modern means to add to
sway', *New York Times*, 4 October 2011

Leo G. Sands, *Guide to Mobile Radio* (New York: Greensback
Library, Inc., 1958)

Jeffery Sconce, *Haunted Media: Electronic Presence from
Telegraphy to Television* (Durham: Duke University Press,
2000)

Kim Severson and Robbie Brown, 'Outlawed, cellphones
are thriving in prisons', *New York Times*, 2 January 2011

Taimoor Shah, 'Taliban threatens Afghan cellphone
companies', *New York Times*, 26 February 2008

Daniel Jordan Smith, 'Cellphones, Social Inequality, and
Contemporary Culture in Nigeria', *Canadian Journal of
African Studies/Revue Canadienne des Études Africaines*
40(3) (2006), pp. 496–523

Pelle Snickars and Patrick Vonderau (eds.), *Moving Data:
The iPhone and the Future of Media* (New York: Columbia
University Press, 2012). Edited collection of essays on the
iPhone, from a media studies perspective

Dan Steinbock, *The Nokia Revolution: The Story of an
Extraordinary Company that Transformed an Industry*
(New York: AMACOM, 2001). Business history of the
Finnish giant.

Alan Stone, *How America Got On-Line: Politics, Markets, and the Revolution in Telecommunications* (Armonk: M.E. Sharpe, 1997)

Hiroko Tabuchi, 'Why Japan's cellphones haven't gone global', *New York Times*, 20 July 2009

Times of India, 'Mischievous SMSs on northeast people now doing the rounds in Delhi', 18 August 2012

Matthew Wells, 'New Yorkers fight mobile phone ban', BBC News, 25 May 2006 http://news.bbc.co.uk/1/hi/world/americas/5013424.stm

Charlotte Windle, 'China's rich fuel mobile revolution', BBC News, 5 December 2005 http://news.bbc.co.uk/1/hi/business/4500692.stm

World Bank, *Maximising Mobile*, 2012 http://go.worldbank.org/FFOU51MTQ0

Peter Young, *Person to Person: The International Impact of the Telephone* (Cambridge: Granta Editions, 1991)

Index